Building Better Roads: Iowa's Contribution to Highway Engineering 1904–1974

Leo Landis

Center for Transportation
Research and Education
Iowa State University
Ames, Iowa 50010-8615

Center for Transportation Research and Education, Iowa State University Research Park, 2625 N. Loop Drive, Suite 2100, Ames, Iowa 50010-8615. Marcia Brink, Editor. Design and layout by Tom Hiett and Donna Fincham, Creative Services, Instructional Technology Center, Iowa State University.

Printed in the United States of America.

Landis, Leo
  Building Better Roads: Iowa's Contribution to Highway Engineering 1904–1974
  Selected bibliography: p. 87
  Includes index.

Library of Congress Catalog Card Number: 97-66382

ISBN: 0-9652310-1-1

The author gratefully acknowledges the generous permissions granted by University Archives, Parks Library, Iowa State University, Ames, Iowa, and by the Iowa Department of Transportation Library, Ames, Iowa, to reproduce the photographs used herein.

University Archives: pp. xi, 2, 6, 11 (bottom), 14 (top), 25, 35, 44, 46, 48, 50, 61 (top), 65 (bottom), 66

Iowa DOT Library: pp. cover, i, iii, viii, xii, xiv, 1, 3, 7, 9, 11 (top), 12–13, 14 (bottom), 15–16, 18, 20, 22–23, 27–34, 38, 40–43, 49, 51–54, 57–60, 61 (bottom), 62, 65 (top), 68–71, 74, 82, 86

Support for this project was provided by the Iowa Department of Transportation and by the Center for Transportation Research and Education at Iowa State University. The opinions, findings, and conclusions expressed in this work are those of the author and not necessarily those of the Iowa Department of Transportation.

Table of Contents

List of Tables

Preface

Written in celebration of the 75th anniversary of the Transportation Research Board, *Building Better Roads: Iowa's Contribution to Highway Engineering 1904–1974* documents the history of highway research in Iowa. To frame the work in a larger context, the text refers to related issues important to highway engineering, such as legislative acts and general research. But the primary focus is the development of highway engineering and research and, particularly, Iowans' participation in that development. The Iowans who helped shape highway engineering were investigators, planners, and innovators whose story could have unfolded in a number of ways. The actual story is one of public service, education, and research, a tradition that continues today.

I incurred numerous debts and received many indulgences while writing *Building Better Roads*. My wife, Nancy Landis, exhibited great patience throughout the project. Doug Hurt, my major professor, directed the research and writing, and his oversight was invaluable. Tom Maze, director, and Marcia Brink, editor, of the Center for Transportation Research and Education at Iowa State University provided financial support and editorial expertise, respectively. Ian MacGillivray, director of the Engineering Division at the Iowa Department of Transportation, provided the impetus for this project and financial support from the department.

A number of former Iowa State Highway Commission engineers provided information about Iowa's record in highway engineering. These engineers included Don Anderson, Gerhard "Gus" Anderson, George Calvert, Robert Given, Ray Kassell, Tom McElherne, and Don McLean. Stanley Ring and R. L. Carstens, professors emeritus of Iowa State University, provided insight on historical and ongoing efforts toward improved highways and highway research. George Sisson and Vernon Marks, contemporary Iowa Department of Transportation engineers, contributed their perspective on present and past work conducted by the department and its parent, the Iowa State Highway Commission. A former director of the Iowa Department of Transportation, Warren Dunham, took time to reflect on his tenure at the department, adding valuable context to contemporary events. I appreciate the valuable details these professionals provided about their experience as highway engineers.

Hank Zalatel and Connie Hasellhoff of the Iowa Department of Transportation's library freely provided direction and assistance; their work is a credit to their profession. Martha "Sam" Koehler, Sherri Harris, and Desi Asklof of the department's Records Management office allowed me to use the department's photograph collection, and the department's photographer, Bill Burns, professionally reproduced many of the images in this book. Becky Jordan, Betty

Road sign, c. 1925 (facing page)
Until 1926 Iowa counties were responsible for their own road signs, so there was no uniformity in signage across the state. Note that this sign is in the middle of the road.

Erickson, Tanya Zanish, Tyler Walters, and Glenn McMullen of the Archives and Special Collections Department at Parks Library, Iowa State University, offered suggestions and assistance whenever necessary. I enjoyed working with them.

Finally, three undergraduate students at Iowa State University, Lisa Green, Nitesh Mehrotra, and Jennifer Westfall, tran-scribed the interviews; their dedication and diligence made the use of the inter-views possible.

I would be remiss if I did not acknowl-edge John and Annabelle Landis, my parents, who have allowed me to pursue my interests and have cushioned my disappointments. This book is dedicated to them.

—Leo Landis

Aeronautical and Civil Engineering panel (facing page)
"When Tillage Begins, Other Arts Follow," designed by Grant Wood.
University Art Collection, Iowa State University, Ames, Iowa.

x

Introduction

Americans have always been on the move. Nations of native Americans roamed the continent for centuries before Europeans immigrated and spread across the country. But the ability to travel hundreds of miles a day on quality paved roads is a relatively recent development. The automobile and durable, hard-surfaced roadways are artifacts of 20th century America. By 1930, coordinated research programs regarding all aspects of transportation were bearing fruit in a modern national transportation network.

The professionalization of highway engineering developed nationally during the Progressive movement, about 1900 to 1920. During that era, idealism and the belief that government programs benefited the public permeated American culture. State governments established numerous commissions to address concerns like public health and state parks. As the nation embraced the use of the automobile, the public desired improved roads and looked to government to provide them.

In Iowa, promoting and investigating hard-surfaced roads became a cooperative effort of state government and university officials. State engineers and university professors worked together to create and develop an enviable transportation system and to help form and direct national highway policy.

Building Better Roads: Iowa's Contribution to Highway Engineering 1904–1974 discusses the beginnings of the Iowa State Highway Commission and the response of Iowa engineers to the demand for quality roads. It also documents Iowa's significant, perhaps unequaled, contribution to modern highway engineering. Creative and forward-thinking engineers like Anson Marston and his student Thomas H. MacDonald influenced not only the state but also the national direction of highway engineering from 1900 to 1950.

Of course, the work of Iowa's transportation engineers has not always provided long-term dividends. For example, early experiments with insulation under roadways to prevent freeze-thaw damage caused surface icing problems, and weigh-in-motion studies in the 1960s were unreliable and never overcame technological impediments. Nevertheless, Iowa transportation engineers have always been enthusiastic experimenters who have used their "failures" as the bases for untold successes.

The record demonstrates the excellent return Iowans have received from the hard work and persistence of dedicated, knowledgeable, and innovative state highway officials and their academic research partners. The following pages tell their story.

Road school, 1906 (facing page)
The Iowa State Highway Commission sponsored road schools to train highway workers from all over the state in the latest methods of road construction and maintenance.

Chapter 1

Beginnings: 1904–1920

At the end of the 19th century, the United States stood poised on the edge of a transportation revolution. Although trains had been carrying Americans and their goods across the continent for 50 years, two radically new modes, airplanes and motor vehicles, were about to shape the American landscape. Air transport would remain a luxury through the first half of the 20th century, but by 1920 automobiles would be available to the public on a wide scale, changing forever the public's expectations regarding personal transportation.

Suddenly the only barrier to quick, convenient, relatively inexpensive personal travel was the lack of adequate roads. The country quickly needed more roads and better roads. To take full advantage of automotive transportation, federal and state governments had to modernize the road system.

For the self-sufficient citizens of Iowa, deciding to dedicate funds to highway work and to create a state agency with oversight of the road system did not come easily. Iowans noisily voiced their concerns regarding increased taxation and bureaucracy as well as intrusion by the state into local decision making. Therefore, the first state agency established by the Iowa legislature to oversee the road system was only an advisory and educational agency with limited authority, rather than a separate state department with supervisory duties.

Automobile on muddy highway, c. 1920
Earthen roads easily became a quagmire in any season. Iowa's soil is good for farming but poor for earthen road construction. The systematic investigations of the Iowa State Highway Commission allowed Iowa to develop a modern highway system.

In 1904 the Iowa State Highway Commission was established at Iowa State College (today Iowa State University of Science and Technology) in Ames. From 1904 to 1913 the commission acted primarily as an instructional office within Iowa State College. It sought to convince the public that a state system of improved roads was in their best interest and to determine the most affordable methods of road planning and construction. During the first 15 years of the highway commission, Iowa State College personnel, highway commission engineers, community leaders, and even newspaper editors promoted and criticized the work of the commission. The direction and support provided by these individuals were critical to the commission's early success.

Rutted road, 1919 (facing page)
This photograph was taken just east of present-day Interstate 35 near Ames, Iowa, on the roadway that later became Highway 30, also known as the Lincoln Highway.

From the numerous challenges of road construction in Iowa came many solutions born of creativity, persistence, and scientific investigation. The work of the Iowa State Highway Commission, the Iowa State College Division of Engineering (now the College of Engineering at Iowa State University), and private individuals established a foundation for highway research and construction that would create models for other states and regions. The problems encountered in the creation of a state highway commission, although sometimes unique to Iowa, also had national applications. The contributions of Iowans to highway research and construction provide an excellent case study of interaction among national, state, and local officials.

Iowa State College Engineering Hall, c. 1905
The Iowa legislature authorized the creation of the Iowa State Highway Commission in 1904. The commission resided at Iowa State College and, until reauthorization in 1913, served primarily in an advisory capacity.

Through the efforts of the Iowa State Highway Commission and Iowa State College, a pattern of nationally recognized highway engineering and research developed. The training and experience gained by engineers and entrepreneurs in Iowa served not only the state but the entire nation. Additionally, Iowa engineers promoted and served the national Highway Research Board, especially in the board's critical fledgling years, to develop a national program of highway research. Eventually Iowa established its own highway research board to solve problems directly related to Iowa.

As a result of these and other activities in Iowa during the first decades of this century, any study of the foundations of modern highway engineering must include a discussion of the Iowa experience and of the state's engineers and their contributions.[1]

A highway commission is established

The years 1904 through 1919, a formative era for the Iowa State Highway Commission, can be divided into two periods. During its first decade, the commission provided advice and instruction on highway engineering, but the function of the commission remained limited. When it was established by the Iowa legislature in 1904, the commission had no state budget for highway construction or maintenance, and for many years all Iowa roads outside of cities and towns continued to be under the local jurisdiction of townships. Counties built and maintained county bridges and sometimes assisted townships with road work.

Almost from the beginning, the Iowa State Highway Commission lobbied the Iowa legislature to authorize a more technically directed highway agency independent of Iowa State College, with authority to develop standard highway plans and specifications for the state. In 1913 the state legislature responded by reorganizing the commission, and the Iowa State Highway Commission entered a new era as a separate state agency, although it remained housed on the campus of Iowa State College until the mid-1920s. The 1913 legislation expanded the Iowa State Highway Commission's duties and provided funds to establish a staff of trained engineers and support personnel. In addition, the commission was given general supervisory control over county and township road officials.

The Iowa State Highway Commission reached maturity and stability under chief engineer Thomas H. MacDonald, who joined the commission in 1904 as its only full-time employee and oversaw its limited responsibilities. When he left the commission in 1919, the Iowa State

Highway Commission had 156 employees, an annual payroll of more than $100,000, and oversight responsibilities for the state primary road system.[2]

The central location of the Iowa State Highway Commission in Ames enabled the commission to maintain statewide interests while, for the most part, avoiding political fights with the Iowa legislature in Des Moines. As early as 1909, MacDonald expressed concern that the Iowa State Highway Commission might be moved to Des Moines, an action that he believed would subject the commission to political manipulation. He and other Iowa State Highway Commission engineers suggested that the Ames location allowed the engineers at the commission plenty of time to prepare for emergency meetings at the capitol or to arrange for any visits from state representatives, senators, or the governor, while maintaining the commission's close physical connection to the facilities and resources available at Iowa State College. The Ames location, with its proximity to the college and convenient distance from Des Moines, played an important role in the development of the Iowa State Highway Commission.[3]

Iowa's physiography

To appreciate the various difficulties posed in creating a quality state road system in Iowa, it is necessary to understand the geological and topographical characteristics of the state. The state's landscape and soils result primarily from three glacial drifts. The southwestern third of the state is covered by the Kansan drift. The north-central area is covered by the youngest formation, the Wisconsin drift. In the northeast, the Iowan drift area prevails, except for the area along the Mississippi River, which is driftless. The southeastern counties along the Mississippi River are part of the Illinois drift.

Glacial Drift Sheets of Iowa[4]

Each area presented special problems for highway engineers. The Kansan drift area has deep valleys cut by streams that offer good drainage but complicate road construction. This area contains some deposits of aggregate. The Wisconsin drift area is flat and plagued by poor drainage and peat. The Iowan drift area also has level terrain that makes drainage difficult, but it has the most extensive deposits of gravel for road building. In the pre-concrete era of road building, deposits of stone and gravel served as valuable material for crushed stone (macadam) roads, but access to these materials proved to be a problem for the highway commission.

Before 1919, road construction usually relied on local materials and the natural landscape. Although some Iowa communities had experimented with portland cement concrete and brick roads, earth and macadam served as the primary road surfaces during this period. Earthen roads often were little more than graded stretches of ground, cleared of weeds and grass. Earthen road soils are grouped into two basic types, clay and loam. Clay soils offer the potential for an acceptable surface, but present numerous problems. If the surface does not compact properly, it becomes soaked with water, and mud

3

collects on the tires of vehicles. Moreover, clay roads can not be graded easily. Loam roads are made from soils that are porous and granular in nature. These roads can be graded more easily than clay roads and can be drained with tile. Macadam roads consist of crushed stone, especially limestone or granite, bound together by water and stone dust to create a harder surface.

In the first two decades of the 20th century, road builders in all regions of Iowa usually built earthen roads or responded to the public's desire for the improved macadam roads. Creating a road system from these materials—a system covering more than 55,000 square miles and linking the rural and urban areas of the state—would require expertise, dedication, and creativity.[5]

Leadership

Based on a plethora of achievements, the engineers and highway professionals of Iowa justifiably claim an honored spot in the fields of highway engineering, research, and construction. Iowa State Highway Commission leaders like Thomas H. MacDonald and Fred White provided direction and leadership for the agency in its early years. At Iowa State College, engineers like Anson Marston, first dean of engineering, and Thomas R. Agg, professor of engineering, supported the efforts of the commission in numerous ways.

However, credit is not limited to chief engineers and deans. Employees of the Iowa State Highway Commission and Iowa State College developed materials, machinery, and procedures that benefited the state, nation, and world. The significance of Iowa's contribution has not gone unrecognized, and Iowa engineers have received many national awards. Occasionally the acknowledgment has been less formal. According to George Calvert, a retired engineer from the Iowa Highway Commission, a contractor from another state once remarked at a national conference, "Don't you think it is time we help Iowa and do some of these things?"

This is not to say that Iowa's contribution is better or more important than the contributions of other regions, but one could not fully appreciate and understand the automobile and motor transport revolution in the United States without examining Iowa's stellar example. Iowa highway engineers offered leadership and provided innovative solutions for difficult problems.[6]

Anson Marston

Out of the American bicycling craze of the late 19th and early 20th centuries came the "good roads movement." The goal of the Iowa Good Roads Association, established in 1903 in Des Moines and consisting of individuals and organizations with vital interests in transportation, was to advocate for better roads through supportive legislation and adequate funding. As automobiles came on the scene, the good roads movement became the impetus for highway research and construction for automotive transport.

Iowa State College Professor Anson Marston sought a role for himself and the college with the creation of a state road agency. Consequently, the history of the Iowa State Highway Commission is inextricably tied to Iowa State University (then Iowa State College). The highway research started at Iowa State College has continued in Iowa, and today all the regent institutions support the work of the Iowa Department of Transportation, which replaced the Iowa State Highway Commission in 1974.

There is no doubt that Anson Marston's vision dramatically shaped the direction of Iowa highway work. He vigorously promoted the creation of the Iowa State

Highway Commission and was a member of the commission from its inception in 1904 until 1927. In 1904, the same year the commission was established, Marston was appointed the first dean of the College of Engineering (then the Division of Engineering) at Iowa State College, an appointment he held until 1932. In his dual role as an Iowa highway commissioner and the dean of engineering at Iowa State College, he provided the impetus for state and national efforts to improve road construction and research. Marston not only engaged in his own research but was instrumental in educating other engineers who would become important to the Iowa State Highway Commission and federal agencies.

But the story almost did not happen that way. Nearsightedness by the Iowa legislature—and a tempting offer from the University of Wisconsin—almost cost Iowa one of its leading transportation engineers. In the critical year 1904, F. E. Turneaure, dean of engineering at the University of Wisconsin, nearly enticed Marston to leave Iowa to work with him in Madison. Here is how events unfolded:

By 1903 Marston understood clearly that before road improvements could be undertaken seriously in Iowa, the public had to support the establishment of a highway commission. That year he sought to establish an engineering experiment station at Iowa State College to begin the necessary research to modernize highway engineering and, by disseminating that research, to publicize the state's great need for an improved road system and a centralized state road commission. Closely allied with him in the effort to establish the experiment station were Charles Curtiss, dean of agriculture at Iowa State College, and Albert B. Storms, the president of the university.

Into this scenario stepped Turneaure, who contacted Marston and offered him a position at the University of Wisconsin in Madison. Convinced that the Iowa legislature would fund the experiment station, Marston initially declined Turneaure's offer in early March 1904. But funding for the engineering experiment station was not allocated, and a disappointed Marston wrote Turneaure expressing regret that he had not accepted the position in Wisconsin.

A flurry of telegrams and correspondence followed as Turneaure renewed his courtship of Marston with inducements like an "opportunity for experimental work and outside work, a larger salary than they pay other Professors of Engineering, and a guarantee of a considerable income in addition from state work of which the University has control." Such opportunities were not easily declined, and Marston accepted a position in sanitation engineering at Madison. Word of his decision spurred a counter campaign by Iowa State College officials to keep him in Iowa, concluding with a personal trip to Madison by President Storms to ask for Marston's release from his commitment at the University of Wisconsin.

Marston now gained funds from the college to establish the Iowa State College Engineering Experiment Station, received a significant pay increase, and was appointed the first dean of engineering at the college. In 1904 the Iowa legislature also established the Iowa State Highway Commission at Iowa State College. If Turneaure had not released Marston from his commitment, it is questionable whether the Iowa State Highway Commission would have been established at Iowa State College, if at all. And if Marston had pursued sanitation engineering, his role in the history of highway engineering would have been minimal.[7]

By 1907 the Iowa State Highway Commission still acted only in a primarily advisory capacity, but Marston articulated his belief in the need to centralize authority in the area of highway engineering. He contended that "disjointed local efforts are no longer sufficient to meet the conditions of the present time." The state administration of roads would eliminate the "enormous waste and inefficiency" that resulted when road administration was controlled at the local level.

Consistent with progressive opinion of the time, Marston believed that state supervision ensured quality and uniform work. The goal for the state, according to Marston, should be to "systematize" state road work according to a three-part plan: systematize highway construction, provide regular funding, and raise funds to construct demonstration roads. An optimistic Marston believed that if such a plan were instigated, "Iowa roads, as well as Iowa agriculture, will in the future be found in the first rank." In other words, allocating centralized highway authority to the state highway commission would solve problems and benefit the general welfare of the state.[8]

Marston was convinced that centralized administration of Iowa's highways would not only prove more efficient and economical than local administration but also would reduce or eliminate graft

Anson Marston
Marston acted as a catalyst for highway engineering in Iowa. He served as a commissioner from 1904 to 1927. In 1919 Marston delivered a speech to the American Association of State Highway Officials (AASHO) to promote nationally integrated highway research.

On April 6, 1904, F. E. Turneaure, dean of engineering at the University of Wisconsin-Madison, wrote Iowa State College Professor Anson Marston, "I am delighted to know that you will come with us. I shall feel now that one of our most important departments is in safe hands." After weeks of protracted correspondence, Turneaure believed he had finally lured Marston from Iowa State College. Turneaure concluded his note, "I feel like apologizing to the state of Iowa for robbing her of so good a man, but at the same time am too well pleased to feel very blue about it."

By the end of April 1904, the apology was not necessary. Marston had received numerous counter enticements from President Albert B. Storms of Iowa State College and chose to remain in Ames.

No field of study at the college or, for that matter, in the state of Iowa reaped the benefits of Marston's decision more than highway engineering. He became Iowa's guiding force for better roads and professionalization in road construction, and his influence in highway engineering shaped the state and nation. Marston directed training and research at Iowa State College and, under his direction, the Iowa State Highway Commission became a recognized leader in highway engineering. Marston's example of leadership, educational extension, and excellence in research has been followed by scores of Iowa highway engineers, providing the state with a legacy of accomplishment. These accomplishments were not inevitable but resulted from effort and innovation in the face of changing transportation needs.

Marston remained at Iowa State College and continued to influence students as a professor, dean, and dean emeritus. He would finish his career at the institution, dying, ironically, in an automobile accident in 1949.

and preferential treatment of contractors. Together with the highway commission, Marston characterized the sometimes cozy relationship between county boards of supervisors and contractors as the "old system" of highway construction. The commission applied this term especially to bridge and culvert construction.

In 1913 the state legislature did reorganize the Iowa State Highway Commission, making it an agency separate from the university and giving it more centralized authority over the state road system. With increased oversight by the Iowa State Highway Commission, a new system of highway engineering soon replaced the inefficient methods of the past and provided Iowans a better infrastructure.[9]

Born in Illinois in 1864, Marston studied engineering at Berea College and received his degree from Cornell (New York) University in 1889. He worked for a branch of the Illinois Central Railroad and later the Missouri Pacific Railroad. In 1894 Marston received an appointment to Iowa State College to teach civil engineering. Among other courses, he taught Roads and Pavements, a class required for second-semester senior civil engineering students. The course covered road materials, road construction, and costs of roads and pavements, and Marston's teaching drew praise from many students and colleagues.

Some time after his appointment as dean, the civil engineering students composed a song, sung to the tune of the Doxology, with the following lyrics:

> All hail our dean with joyful shout,
> He helps us drain our cellars out;
> King road drags are his chief delight,
> His water tanks are water tight.
> Macadam roads you sure can make,
> If Roads and Pavements you will take,
> For consultation he is great;
> We'll all agree our dean is up to date.
> Amen.[10]

The Roads and Pavements course allowed Marston to instruct many students who would later become well-trained highway engineers at the Iowa State Highway Commission. Indeed, the significance of Marston's teaching is great. The first two chief engineers at the commission, for

Man and boy dragging road, c. 1915
Earthen roads demanded constant attention, and local officials and farmers used the road drag to maintain a usable road surface. Legislation in 1913 created township maintenance districts, and workers maintained earthen roads with shovels and road drags. By 1919 one county engineer believed the road drag's time had passed, but many Iowans still traveled earthen roads.

Iowa Highway Commission Engineers 1913[11]

Name	Position	College, Year of Degree
MacDonald, Thomas H.	Chief Engineer	Iowa State College, 1904
Kirkham, J. E.	Consulting Bridge Engineer	Missouri, No Date
McCullough, C. B.	Assistant Engineer	Iowa State College, 1910
Ames, J. H.	Assistant Engineer	Iowa State College, 1911
White, Fred	Assistant Engineer	Iowa State College, 1907

example, whose service spanned 1904 to 1919 and 1919 to 1952, were students of Marston. In 1913 four of the five engineers on the commission had received their degrees from Iowa State College. The staff also included a drafts-man, James A. Paulson (Iowa State College 1912), and two stenographers, Annie Laurie Bowen and Merle Crabtree. Marston's influence continued at least through the first two decades of the commission's existence. In 1918, at least 15 of 37 engineers and commission employees had received engineering degrees from Iowa State College.

Marston can rightfully be called the catalyst for modern highway construction and research in both Iowa and the nation. He instructed numerous students in his Roads and Pavements course and inspired further investigation by his students. He accepted advice from both peers and students. He sought the best existing models as part of his work with the highway commission. He traveled to New Jersey, New York, and Massachusetts to examine the condition of road work and highway commissions in the east. New Jersey and Massachusetts led the nation in the movement to establish state agencies to promote better roads, and Marston wanted to learn from their experiences.

Thomas H. MacDonald

Anson Marston's equal in the history of Iowa highway engineering is surely Thomas H. MacDonald. A student of Marston's, MacDonald would later become the first chief highway engineer for the state of Iowa. MacDonald's family moved to Montezuma, Iowa, from Colo-rado when he was three. He received his civil engineering degree from Iowa State College in 1904, having written his senior thesis on the Good Roads problem in Iowa. That summer he joined the Iowa State Highway Commission as the assis-tant in charge of the Good Roads investi-gation. Later, as chief engineer,

MacDonald administered the state road program and led the highway commission through its formative years. MacDonald served the highway commission for almost 15 years until 1919, when he was appointed commissioner of the Bureau of Public Roads in the U.S. Department of Agriculture.

As chief engineer at the Iowa State Highway Commission, MacDonald selected the district engineers and repre-sented the engineers at commission meetings. He also disseminated informa-tion from commission meetings to the engineering staff and county officials. The commission required a diligent and effective chief engineer to provide focus for the agency and to serve as a credible representative to the public, and MacDonald proved an effective leader.

While he was at the Iowa State Highway Commission, MacDonald also served on the staff or council of the Engineering Experiment Station at Iowa State College. For many years the Engineering Experi-ment Station served as the research agency for the commission. Together, Marston and MacDonald directed Engi-neering Experiment Station projects, and MacDonald established a precedent for effective management and credibility of the station's work.

When MacDonald became the first chief engineer of the Iowa State Highway Commission, the commission was still technically a part of Iowa State College. By 1909 MacDonald realized the impor-tance of establishing the commission as a completely separate entity. He expressed discontent with the university's practice of referring to the highway commission as the Good Roads Department. MacDonald also wanted a separate environment with less interruption. He warned, however, that when the commission became a separate entity, "there will be a deter-mined effort to remove the work from the college to Des Moines."[12]

Early research efforts

Under Marston's direction, the Iowa State Highway Commission followed issues of national interest and began to take part in national research. Marston was then able to use his national reputation to influence research on the state level.

Collecting data

Before the Engineering Experiment Station was established at Iowa State College, student thesis research provided information on highway matters. In his position as professor of civil engineering and later as dean, Anson Marston encouraged his students to engage in research related to highway engineering.

Two students completed a traction study in 1903, and a year later Allen B. Chattin and Ray McClure submitted a thesis titled "Good Roads Investigations." This study documented the tractive resistance of various Story and Polk county roads and offered many conclusions, including the undocumented claim that the average farmer traveled to market three to four times a week. Chattin and McClure expressed surprise at the frequency of such trips and asserted that the best and most economical solution for the care of roads would be the appointment of a competent engineer in each county to oversee county roads. This was probably the first instance of Marston's students not only engaging in scientific research but also using research results to suggest changes in public policy.[13]

In 1904 L. T. Gaylord and Thomas H. MacDonald submitted a senior thesis titled "Iowa Good Roads Investigations." MacDonald and Gaylord sought to replicate the conditions encountered by Iowa farmers. They studied eight roads in the Ames area and 14 routes in Linn County using a team of horses, a typical farm wagon, and a ton of sand. To measure the draft required to pull the load, the men fastened a dynamometer between the team and wagon that plotted the necessary draft power on a piece of paper. The men tried to choose typical roads. In the Linn County test, the men employed an "old resident" with road building experience as a driver. Based on the collected data, MacDonald and Gaylord asserted that macadam road surfaces required the least draft. The amount of power required to pull a load on dirt roads could be seven times greater than the draft necessary on broken-stone roads.

Not only did they report the data gathered from the road tests, MacDonald and Gaylord also provided commentary and suggestions regarding Iowa highways. They freely admitted the necessity of gaining public support through discourse and education to further the cause for better roads: efforts to create quality roads would "remain spasmodic and discontinued until a State Highway Commission is established," they reported. Their essay argued that such a commission should have the power to construct roads funded, in part, by state money. MacDonald, possibly influenced by Marston, had decided that state support and oversight would guarantee quality highways. A final recommendation urged better drainage to improve poor roads "on a systematic basis in each county." With little equivocation, this paper established the Marston and MacDonald plan for improved roads.[14]

Marston continued to use his students and former students to generate highway data. One of the earliest programs that

Thomas H. MacDonald
MacDonald joined the Iowa State Highway Commission in 1904 and served as its first chief engineer. In 1919 he received an appointment as chief of the Bureau of Public Roads, a position he held until 1952.

9

Marston and the Iowa State Highway Commission advocated included a state road census conceived by Chattin. Chattin suggested that Marston lobby to have a road census conducted as part of the Iowa Agricultural Census of 1905. This state-wide survey would produce data on road use similar to data collected by Chattin and McClure in their senior thesis. The data would be published in an Engineering Experiment Station bulletin and provide an economic rationale for promoting improved rural roads. Although the enumerators collected the data, the state did not allocate the necessary resources to tabulate the results.[15]

Engineering Experiment Station

Between 1904 and 1919 the Engineering Experiment Station at Iowa State College served as the primary data gathering agency and clearinghouse for the Iowa State College engineering research laboratories concerning highway engineering and construction for the state highway commission. Anson Marston served as its first director and continued to use student researchers extensively.

With Marston's leadership, the Engineering Experiment Station researched highway engineering problems and published bulletins related to concrete in 1904, limestones in 1907, and gravels in 1916. Marston first published his research on the theories on loads for pipes and culverts in a 1914 Engineering Experiment Station bulletin. These writings provided the basis for modern load theory for conduits used by the American Society for Testing and Materials.[16]

The Engineering Experiment Station also began to investigate the critical area of materials testing and to determine the location of suitable aggregate. At that time macadam (crushed stone surfaced) roads received the greatest support from the commission for road building. However, limestone suitable for road building

existed only in the eastern part of the state. These aggregate investigations became important during the early 20th century.[17]

Although the size of the Engineering Experiment Station staff during its early years is unknown, by 1913 20 engineers served the station. In addition to Marston, other members of the staff included MacDonald and Roy Crum. Thomas R. Agg soon joined the staff as a road engineer. The station created a Good Roads section to study highway construction projects. Specifically, the Engineering Experiment Station served as a scientific testing agency that disseminated pure and applied scientific information relative to highway engineering. Marston's students and the Iowa State College research laboratories served the Iowa State Highway Commission well in the years ahead.

Bulletins

The earliest Engineering Experiment Station bulletin related to highway engineering, published in June 1905, was titled "The Good Roads Problem in Iowa," and it summarized the road difficulties in the state. The authors, probably Anson Marston and Thomas H. MacDonald, used data developed from the research conducted by Allen B. Chattin and Ray McClure, linked the prices of farm commodities to road conditions, and noted that when roads became impassable the prices of commodities increased. The bulletin also argued that to ensure "social and educational advantages," quality roads must be constructed, and it provided a plan that discussed the value of a state-managed highway system.

By 1914 the Iowa State Highway Commission had its own instrument for disseminating information, the *Service Bulletin.* Through the *Service Bulletin,* the commission published short essays regarding commission work and highway research results produced by both the commission and the Engineering Experiment Station. In two consecutive issues, Roy Crum published articles on the grading and testing of concrete. These reports analyzed concrete components and suggested standardized tests to arrive at better mixes. Crum's research studied the size and character of aggregate, as well as the identification of natural impurities in the aggregate. This seminal research indicated the transition of concern from earthen roadways to more durable highways because concrete roads could withstand the increased traffic of automobiles. The Iowa State Highway Commission and Engineering Experiment Station performed the necessary research to prepare for this new direction.[18]

Education and standardization

As the highway commission matured, it began to define its own objectives rather than rely strictly on the general legislative mandate. Admittedly, the commission had been chartered and could only act within the parameters of the legislation that created it, but the commission began to explicitly state how it would function based on the legislation. The commission's annual report for 1906 listed six functions for the commission, which included investigation, experiments, school of instruction for road officers, plans and publications, demonstration work, and road meetings. The commission also began to document the condition of roads with photographs, test concrete bridges and culverts, test materials in a laboratory, and conduct demonstrations.[19]

To establish its legitimacy as a useful authority, the original Iowa State Highway Commission initiated outreach programs and standard plans for construction. The commissioners and engineers traveled the state promoting better roads and the need for a centralized, state authority on highways. The commission also published a manual with standard plans for road construction and continued to promote a system of highway construction.[20]

Demonstration trains and road schools

The highway commissioners believed competency came through standardization and training. In 1905 demonstration trains sponsored by the commission and Iowa State College served as one method to promote better roads on a local level and educate the public about the care and upkeep of roads. The trains were traveling educational programs. Thomas H. MacDonald and other experts traveled the state by rail, stopping in towns and cities to provide practical demonstrations of earthen road care.

In April 1905 the Chicago & North Western Railway sponsored a Good Roads demonstration train that traveled from west to east from Onawa to DeWitt. Not to be outdone, the Burlington Railroad sponsored a train in the fall that stopped at 16 towns between Council Bluffs and Burlington.

The primary goal of these trains was to promote a device called the King road drag. Promoter D. Ward King of Missouri traveled on the trains to instruct local citizens on the construction and use of his drag. Using the drag on earthen roads

Concrete pipe test, 1914
The Iowa State College Engineering Experiment Station and the Iowa State Highway Commission developed standards for testing pipes and culverts. Pipes, culverts, and tiles removed water from the roadway and subgrade and needed to withstand the forces of earth and traffic.

Material Testing Laboratory, c. 1915
The Iowa State College Engineering Experiment Station conducted tests for the Iowa State Highway Commission. Many Iowa State College graduates worked for both the station and the commission.

Road train, 1905
Road trains across northern and southern Iowa were a coordinated effort among private citizens, newspaper publishers, and the Iowa State Highway Commission. D. Ward King, a Missouri farmer, advocated the King road drag for earthen road maintenance. The highway commission used these public events to promote the new state agency. Chief Engineer Thomas H. MacDonald lectured on culverts and drainage.

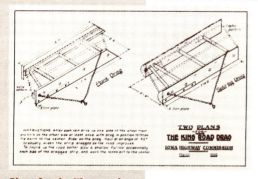

Plans for the King road drag, 1906
Farmers and road officers could build a King road drag based on plans in the highway commission's Manual for Highway Officers published in 1905. The drag could be made of split logs or milled lumber.

proved valuable for creating a crowned and hardened surface that provided better drainage and a smoother surface. The drag consisted of two 10- or 12-foot half logs or planks secured parallel to each other by iron straps and pulled behind a horse. After a rain, a farmer or road officer could drag local roads, filling the ruts and enabling the surface to shed water better during the next rain. If allowed to puddle, the road surface would deteriorate to muck and become impassable. As long as earthen roads were common, the road drag served as a valuable tool for road maintenance.[21]

Besides advocating the road drag, the demonstration trains offered Thomas H. MacDonald an opportunity to promote the Iowa State Highway Commission to the public and place the commission in the public eye as a positive agent for road improvement. Still, the role of the Iowa State Highway Commission on the spring 1905 trip proved negligible. King acted as the primary speaker, and newspaper accounts made little mention of MacDonald: in northwestern Iowa, the

Rolfe Reveille reported that MacDonald addressed the crowd but did not mention the topic of his speech.

On the southern trip in the fall, however, MacDonald received more press. The *Osceola Democrat,* for example, described MacDonald: "[T]hough a young man, he is a civil engineer of skill and great ability." MacDonald lectured on bridges and culverts and advocated the use of concrete culverts, which were proving to be more durable and economical than wooden structures. Overall, the Iowa State Highway Commission received favorable coverage during the spring and fall excursions, and the trains established the press as an ally and agent for promoting good roads.[22]

Besides the demonstration train tours of 1905, highway commission personnel attended farmers' institutes. The commission report for 1907 and 1908 listed 16 institutes and meetings attended by its representatives. The commission also attended other "special road meetings" to instruct the public about proper road construction. Advocacy for the role of the Iowa State Highway Commission was giving way to increasingly important educational efforts. Although the commission report may have overstated the passing of the period of self-promotion, the goal of educating road builders became at least equal to the goal of convincing the public that a highway commission served their interests. The commission contended that it had established itself as a legitimate state agency by 1908. This did not mean that the commission would not face political or professional attacks in the future.[23]

In addition to off-campus road programs like the demonstration trains, the highway commission sponsored road schools in Ames. The first program was a three-day event in June 1905 on the campus of Iowa State College. This road school focused on the Iowa State Highway Commission, field work, construction and maintenance of earthen roads, and machinery, with particular attention to the King road drag. Primary presenters included Anson Marston, Charles Curtiss, and Thomas H. MacDonald, with other experts available as well. The road school continued through 1908 and expanded to include officials from other state highway commissions, such as Missouri.

To better serve the whole state, the commission also held the road school in selected communities outside Story County. Commission employees traveled west in 1907, for example, to hold a road school in Council Bluffs. As part of this workshop, the commission sought to aid in the development of quality roads to link rural services. W. R. Spillman, a postal official in the division of Rural Free Delivery, spoke at the Council Bluffs school on "The Road Problems as Affecting the Extension and Continuance of R.F.D. Service."

In August 1908 the highway commission offered a school in Waterloo. The Waterloo program included the stalwart, King, but also national figures. Secretary of Agriculture James Wilson delivered an address regarding agriculture and quality roads. Officials from the Office of Public Roads and the Missouri highway commission also appeared on the schedule for the five-day workshop.

These practical road school seminars ceased for a time, but the college resumed the road school in 1913 with the reorganized highway commission. Road schools continued through the earthen road period, with a short course held each winter or spring. The workshops provided direct contact and practical experience for county engineers and other highway officials. These programs encouraged interaction and cooperation among the public, county officials, and the highway commission.[24]

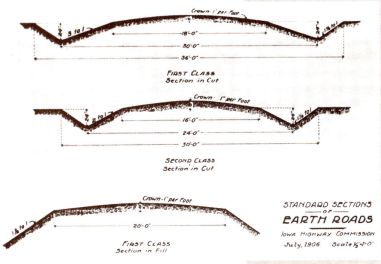

Road cross section, 1906
The Iowa State Highway Commission classified earthen roads based on traffic volume. Each type of roadway had different standards.

Reorganization of the Iowa State Highway Commission in 1913 provided new authority to the commission. The commission's outreach work continued in 1913 and 1914, and its engineers made goodwill tours and presented lectures to community groups. J. S. Dodds, engineer in charge of the commission's education department, participated in 53 programs between November 1913 and November 1914. Dodds delivered lectures to farmers' institutes in six counties ranging from Sac in the northwest to Mahaska in the southeast. He also assisted at Good Road

Children on a properly crowned road, c. 1910
A proper crown on earthen roads allowed the highway to shed water. These children are shown on a properly crowned road.

John S. Dodds (far right), c. 1917
Dodds served the Iowa State Highway Commission and Iowa State College to promote better road building. In the 1910s he traveled to farmers' institutes and fairs to promote the work of the commission.

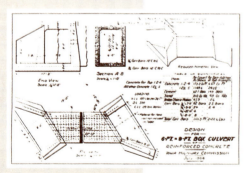

Box culvert plan, 1906
By 1905 the Iowa State Highway Commission had established a system for highway maintenance and construction. The Manual for Highway Officers provided information and plans like this one for box culverts to standardize Iowa highway work.

days, such as that in Calhoun County, and he coordinated the highway commission exhibit at the Iowa State Fair.

Dodds visited the most towns, but Dean Marston, Chief Engineer MacDonald, and Commissioners H. C. Beard and James W. Holden, among others, also engaged in speaking tours to promote the highway commission. The personnel of the commission knew that the public had not completely accepted a state highway agency, and they realized the necessity of continuing to promote their work.[25]

Marston, in particular, aided the outreach activities of the Iowa State Highway Commission. He believed drainage to be of primary importance in highway engineering and continued to test different pipes and tiles for drainage and to promote standardized box culverts based on highway commission designs.

In 1914 the double priorities of solving drainage problems and fostering the

outreach mission of the commission were emphasized at a meeting of local people and highway commission officials in northeastern Iowa. W. D. Fillmore, editor of the *Dows Advocate,* organized the meeting and, along with other editors, promoted the commission's work.

Throughout the existence of the highway commission, friendly editors supported its work. An essay on the Dows trip printed in the commission's *Service Bulletin* further emphasized the importance and the quality of work by the highway commission, as opposed to that of unsupervised and untrained local officials. The bulletin article contrasted a photograph of a box culvert based on a highway commission design to a crumbling culvert constructed under the old system. The implication could not be missed; for economical and quality work, the highway commission provided the necessary expertise and guidance.

Publications
In 1905 the commission took an important step by publishing a standard guide called the *Manual for Highway Officers* to educate county and township road officials. Revised in 1906, the guide summarized the current road conditions in Iowa, described the state's topography, and noted the legislation related to road construction. This manual provided a detailed reference for road construction along with recommendations for county officials. It also advocated the use of the road drag, provided plans for culverts, and presented the results of tests related to materials and tractive resistance.[26]

Other efforts by the commission to educate engineers and the citizenry came in the form of bulletins and state fair exhibits. The *Service Bulletin* promoted the goals of the highway commission and circulated information ranging from the mundane to the sensational. Much of the information related to the commissioners' current projects, long-range plans, safety issues, and the efforts of commission personnel and county engineers.

Articles detailing inadequate bridges that collapsed from the weight of steam engines filled the midsummer issues during the threshing season. In southwest Iowa, three Page County men died between March 1913 and March 1914 when their threshing rigs crashed through inadequate bridges. Such dramatic incidents highlighted the need for commission supervision of bridge design and justified the standard bridge plans used by the agency.

Standards and patent-breaking

With its reorganization in 1913 the Iowa State Highway Commission was authorized to prepare standard specifications and plans for state and local roads and bridges. The commissioners believed competency would be developed through standardization as well as training.

To support its standardized plans, the commission occasionally engaged in litigation. One area of controversy was the commission's bridge designs, which used techniques for which private firms had received patents. The commissioners believed these construction methods belonged in the public domain and, with state support, challenged the patents. This effort began in 1913 when the state legislature passed a bill authorizing the attorney general to act on issues of highway, bridge, and culvert construction. The attorney general's office prosecuted cases and often defended the Iowa State Highway Commission or local officials when patent holders sued them or their contractors for using protected designs. C. B. McCullough, an engineer at the highway commission, dominated the movement and helped to invalidate over 100 patents.

As part of its effort to combat patented bridge designs, the Iowa State Highway Commission sought to make an example of the Luten Engineering Company of Indianapolis. The Luten company held patents for bridges manufactured of reinforced concrete. The commission objected to the Luten bridges for three reasons: the engineers believed the patents were invalid, the bridges were overpriced, and Luten bridges did not meet the commission's bridge standards. The commission used the collapse of a Luten bridge over Squaw Creek in Ames on June 28, 1918, to emphasize the need for its expertise in highway engineering and the commission's dedication to public service rather than profit.

Iowa State Highway Commission state fair display, c. 1930
The highway commission has always promoted its work to the public. County and state fair displays demonstrated the commission's dedication to keeping the public informed about its work. This service allowed the commission to develop support for increased funding for the highway system.

The city of Ames had paid $8,000 for the construction of the Luten bridge, which lasted only 10 years. The highway commission insisted it could have built a more durable structure for $11,000. The bridge's dramatic collapse made excellent press for the highway commission,

Traction steam engine on wooden bridge, c. 1915
Threshing crews moving steam engines and separators often broke through wooden bridges. The Iowa State Highway Commission promoted standardized plans to maintain service and protect the traveling public.

15

Collapse of a bridge on the Lincoln Highway, June 1918
This disaster gave the Iowa State Highway Commission an opportunity to promote the superior quality of commission-designed bridges. Fortunately, a family traveling across the bridge when it collapsed was not injured.

Professionalization of the county engineer

The Iowa State Highway Commission addressed other county and local concerns. Particularly, the commission sought to professionalize the position of county engineer. While some county engineers had professional training, others received the responsibility solely to meet the state requirement that a county appoint an engineer. The county boards of supervisors appointed the county engineers, but the commission had general supervision of county and township road officials and, after the 1913 reorganization of the commission, oversight authority for county engineers and some county roads. In this function, the commission could reward and promote quality engineers or remove those who did not meet the increasingly technical demands of highway engineering.

because a family in an automobile had been on the bridge at the time. The family survived, but the accident helped promote the highway commission's proposition that building bridges privately, without state oversight, was a dangerous practice.

Although the commission made a good argument, the fact that a number of Luten bridges remain in existence today shows that the highway commission may have

In June 1913 Claude Coykendall, a highway commission district engineer, described the Clarke County engineer in southern Iowa as "a reformed scrub carpenter." Coykendall's report described the construction work contracted by the county as inferior and not in accordance with commission guidelines and claimed the county engineer's knowledge was insufficient. Coykendall reported that the county engineer knew only the rudiments of engineering, and that he "would not be a good man for the place, as he antagonizes every one with whom he comes in touch." The Clarke County engineer's only redeeming feature, as far as the board of supervisors was concerned, Coykendall wrote, "is that he will work for $4.00 per day." In spite of the engineer's incompetence and noncompliance with commission guidelines, the overall conditions of the county roads appeared better than average.

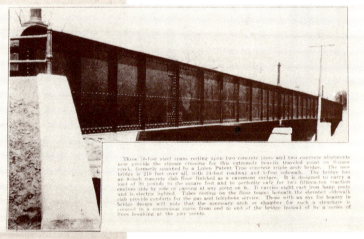

Replacement bridge on Lincoln Highway, 1922
The Iowa State Highway Commission approved the design of the replacement bridge completed in 1922. The bridge was 210 feet long, and the commission boasted that it could carry two 15-ton traction engines simultaneously.

exaggerated a one-time incident to promote the agency. In a number of cases, the state won judgments, or cases against the commission were dismissed, allowing construction of standardized commission plans without fear of legal repercussions.[27]

Nevertheless, when the commissioners met in June 1913 they voted unanimously to have Chief Engineer MacDonald "inform the board of supervisors of Clarke that the services of Mr. Andrews [the engineer] are to be dispensed with immediately and a competent engineer able to do the work employed." In the interim, based on Coykendall's lack of confidence in the situation, he reported it would be necessary to forward the standard commission plans to Clarke County contractors to ensure their work would meet the commission's requirements. The commission sought to enforce its policies and establish professional engineers, even if county roads met common standards.[28]

While the commission replaced incompetent engineers, it emphasized the good work of dedicated county engineers. One of the most sensational articles in the *Service Bulletin* described the death of Mahaska County Engineer A. E. Rommel and illustrated the devotion of county officials. Rommel perished on February 26, 1916, as he and Mahaska County citizen Charles Thomas navigated the flood waters of the Des Moines River in an attempt to dynamite an ice jam threatening the bridges along the river. Rommel and Thomas had completed a number of trips and fired charges to no avail. Their final trip ended in tragedy when the men were tossed from the boat. Thomas climbed a tree to safety, but Rommel, weighed down by a heavy coat, remained in the water. Although he reached a pile of driftwood, Rommel went into shock, slipped into the current, and drowned. The commission used this tragic incident to illustrate the dedication of county engineers to the public good.[29]

The *Service Bulletin* also illustrated the quality work completed by the highway commission. These accounts, though partisan, demonstrated the professional high regard for the quality of Iowa highway work. In 1917 Henry C. Ostermann, field secretary for the Lincoln Highway Association, commended the state for the character and quality of its bridges. He specifically mentioned the practical construction, durability, and weight capacity of Iowa bridges. Ostermann further commented that the state had excellent roads, at least in good weather.[30]

Emphasis on safety

Economy and safety of roads have always been primary goals in the construction of highways. The highway commission emphasized these points in 1915 when it addressed the improper use of roads in its annual report. Particularly disturbing, according to the report, were the cross-state road races that promoted high speeds and a disregard for the "laws of safety" and gave "the wrong impression of the safe uses that can be made of ordinary dirt roads." This kind of reckless disregard for safety caused most accidents. The report also concluded that earthen roads could not be quickly improved. Iowa still had a reputation as a quagmire in the spring.[31]

Federal funding begins

As early as 1914 federal monies were devoted to hard-surfaced roads in Boone and Story counties, and Dubuque County received funds for a road from Dubuque to Dyersville. But it was not until the landmark Federal-Aid Road Act of 1916 that a process was initiated for systematically allocating federal funding to the states for road construction. The federal funds would match state appropriations up to $10,000 per mile. The Federal-Aid Road Act fund initially allowed $146,200 for Iowa, which was increased to $731,000 in 1920.

On July 4, 1917, the federal government inaugurated the process for federally approved and assisted construction of hard-surfaced roads. Iowa submitted

Iowa's first federal-aid project
Iowa's first project under the 1916 Federal-Aid Road Act was a road between Mason City and Clear Lake. This project turned a 10-mile earthen road into a section of first-class portland cement concrete highway.

First federal-aid project completed
By 1919 workers completed Federal-Aid Project 1, and the highway was opened to traffic.

A partnership for research into the next decades

The association of the Iowa State Highway Commission with Iowa State College provided an atmosphere conducive to research and investigation with state and federal support. The college's continued involvement in highway research after the commission's reorganization in 1913 is clear from a report in that year that announced, "After the establishment of the Commission as a separate state department, the Good Roads Experimentation appropriation was used by the college to build up a complete testing laboratory."

Established in 1904 to foster investigation of materials, the Iowa State College Engineering Experiment Station proved its value as a research agency in the first 15 years of the highway commission. This work continued through the second decade of the century in the laboratories of Iowa State College and was conducted largely by members of experiment station staff. At first the college bore most expenses with funds appropriated by the state. As the benefits of the research and testing at the station were realized, counties also contributed funds. In 1917 the college furnished staff, materials, and equipment either free or at a nominal rate, with some expenses charged to counties for oil tests across the state. The Iowa State Highway Commission supported additional research in the second decade of the century.

12 projects, eight of which were approved by the Office of Public Roads in the U.S. Department of Agriculture, and in the fall of 1917 a new era in road work began in Iowa as the state began its first federal-aid hard-surfaced road project funded by the Federal-Aid Road Act.

Northern Iowa received the first benefits of these federal funds. Monies were allocated to construct a 10-mile, type A, 16-foot wide, portland cement concrete section of road between Mason City and Clear Lake, near what is now Highway 18. The contractor completed about one and a half miles before winter prevented further work. The commission's annual report for 1919 highlighted the completion of the road; in the end, federal dollars paid for four and a half miles of this highway. [32]

Summary

By 1919 the Iowa State Highway Commission had firmly established itself as a credible state agency in charge of highway research and construction. The commission had become independent of Iowa State College (although it would physically remain on campus until 1924), had received increased state and some federal support, and had begun to supervise and approve the work in counties and municipalities.

The leadership and facilities to create a modern road system were in place. Strong administration from Anson Marston and Thomas H. MacDonald enabled the commission to receive support from critical private enterprises and the public. Professionally trained highway engineers with practical experience represented the commission across the state, and their proficiency paid dividends. Iowa was prepared to reap the benefits of federally funded, hard-surfaced road construction.

This conscientious public service and professional expertise placed Iowa highway researchers and engineers at the top of the profession. Their experience on the state level drew national attention that would be recognized in the post-World War I era.

Chapter 2

Serving the State and Nation: 1920–1939

On April 7, 1919, Thomas H. MacDonald, recently appointed chief of the Bureau of Public Roads, issued a memo to the staff of the Iowa State Highway Commission announcing, "Effective this date, Mr. F. R. White will assume all executive duties of the Commission." In this note, MacDonald expressed his optimism about the progress of the highway commission and requested that White be given the same support and cooperation that he had received.

The year 1919 proved monumental for the Iowa State Highway Commission, if for no other reason than MacDonald departed and Fred White became chief engineer. White's tenure would span longer than any subsequent chief engineer. From 1919 to 1952 he contin-ued the excellent engineering tradition established by MacDonald and Anson Marston. White guided the state highway commission through hard-surfaced paving, a national depression, and a second world war. Under his direction, the Iowa State Highway Commission built a first-rate road system, prepared the state for four-lane highways, and established a state highway research board.[33]

By 1919 the Iowa State Highway Commis-sion had firmly established itself as the state authority for highway work. It had designated a primary road system and received federal funding to complete permanent roads. The commission continued to face assaults from unfriendly political forces but still increased its authority and prestige. In 1929 legislation gave the commission total oversight and management of the state and county road system. In the post-World War I and depression eras, the commission paved more than 5,900 miles of road and graveled a similar number of miles. White directed the commission in this period of growth, while MacDonald, as chief of the Bureau of Public Roads, led the nation in highway research and planning.

In 1919 Iowa passed its first Primary Road Act, establishing a 6,400-mile road system to connect the county seats and market towns of the state. This legislation gave the Iowa State Highway Commission oversight authority for the expenditure of federal highway funds and authorized road construction to enable highways to connect more than 90 percent of Iowa's city and town population. Until 1927 these road improvement plans remained subject to county supervisor approval, although legislation stipulated that all federal road projects receive state supervi-sion from the state highway commission. Iowa's road system would not fall com-pletely under the Iowa State Highway Commission's jurisdiction until 1929.[34]

Leadership and continuity: Fred White

MacDonald's departure from the Iowa State Highway Commission did not leave a void in executive leadership. Fred White followed in the tradition of Marston and MacDonald. He grew up in Iowa and received his degree from Iowa

Dirt road, c. 1915 (facing page)
Before paving became commonplace, planks were placed over sink holes, ruts, and soft spots in the roadway. Shown is Grand Avenue, Ames, Iowa.

Fred White
White held the position of chief engineer at the Iowa State Highway Commission from 1919 to 1952. He directed hard-surface road construction on primary and secondary roads.

State College in 1907. His senior thesis related to bridge engineering, and his first duties with the highway commission concerned bridge construction. White's first tenure with the Iowa State Highway Commission proved brief (1908 to 1910), but he returned in 1911 as a field engineer, a position he held until 1916. He became a road engineer, then assumed the position of chief engineer upon MacDonald's appointment to the Bureau of Public Roads.

As chief engineer, White first concerned himself with the coordination of the new state laws related to primary and secondary roads (the primary road system consisted of federal and state highways managed by the state; the secondary road system was managed by the counties) and the federal legislation providing federal monies for highway construction. The federal program required that the money be expended for construction and that each state comply with three provisions. First, state legislatures had to abide by the provisions of the federal act. Second, each state had to establish a highway department and give it direct supervision of federal road projects. Third, states had to match or exceed the federal allocations. The roads could be of any character, but Iowa's engineers preferred to use portland cement concrete and gravel.[35]

White described the years from 1900 to 1925 as a formative period in highway

engineering. He recognized the early legislative constraints on, and the political implications of, the highway commission's work and knew that commission engineers had to follow legislative guidelines. Then in 1925, state legislation enlarged the responsibilities of the highway commission from advisory to supervisory, establishing the Iowa State Highway Commission as a "contract-making and contract-executing body."

White considered this new state system to be a significant improvement over the previous system of township jurisdiction over roads, a system that, he believed, produced "loose cogs in the machine." If the highway commission did not perform its new responsibilities satisfactorily, the public, still nervous about the loss of local authority in road administration, had the power to influence state legislators to revise the statutes and diminish the authority of the commission.

Besides some local opposition to state control of roads, another barrier to developing a program for hard-surfacing the state's roads was the lack of state-managed road funds. Until the commission received greater command of road funds, counties would continue, in White's words, to "pour sand down a rat hole." White believed that state, or highway commission, control meant that primary roads would be improved to uniform standards of maintenance. Only if the commission used its increased authority to improve road conditions until they reached a level of relative equity would monies for paving be allotted on a need basis. White tied statewide improvement of roads directly to the increased authority of the Iowa State Highway Commission; without state direction, improvements would occur sporadically across the state.[36]

In the 1920s the Iowa State Highway Commission began to articulate road plans for multiyear periods. One plan proposed by Chief Engineer White was published in the July–August 1923 *Service Bulletin*. It established little more than general goals but created an enduring policy. Covering the years 1926 to 1928, the plan offered details for financing projects in the primary road system, recommended materials for surfaces, and urged completion of grading and draining the primary road system.

The commission especially wanted to develop a statewide road system built with federal, state, and county funds. The *Service Bulletin* article announcing the principles of the plan asserted that the commission sought "[t]o complete within three years a network of highways extending all over the state, connecting every county seat in the state, and forming several routes east and west and north and south which are fully improved, entirely across the state, by grading, draining, bridging, and graveling." Perhaps most important, the plan detailed the commission's goal to construct at least one continuous graded, drained, and surfaced road across the state.

The report also listed 12 Iowa counties believed to present the most difficulties for constructing the state's network of highways. The counties ranged across the entire state, from Cass in the west, to Bremer in the northeast, to Lee in the southeast. The northwest section of state was not included in the list. The primary problems in the 12 counties were topography and the high cost of grading and building bridges.[37]

While White was leading the Iowa State Highway Commission to new goals, Anson Marston was undergoing pressure from the state legislature. Genius and dedication are not always appreciated at

Ditching and grading the roadway, c. 1920
The first steps in creating a highway are ditching and grading. The Iowa landscape challenged the Iowa State Highway Commission as contractors created suitable roadbeds.

home, and such was sometimes the case with Marston. In the 1920s partisan politics led to various attempts in the Iowa legislature to remove Marston from the highway commission. In 1925 the senate passed the Busser bill that would have removed Marston from the commission, but the bill failed in the house. One observer remarked that the removal of Marston would have been "boyish destructiveness," and "beyond sympathy," because of the loss of his expertise to the state.

Marston weathered the storm and remained on the commission until 1927. At that time, he took a leave of absence from Iowa State College to participate in a drainage survey in the Florida Everglades. When he retired from the Iowa State Highway Commission, Marston left a legacy of professionalism and dedication to highway engineering. The state senate, which two years earlier had attempted to oust him from the highway commission, now honored him with a resolution stating that "as a man and an engineer he again leads us to assert, 'Of all that is good Iowa affords the best.'"

Marston believed he left the highway commission in capable hands. In a letter to White he expressed the following statement: "My heart will always be with you and your work with the Highway Commission and its work, and I feel very confident of the continued success of this great public service." Although Marston had concluded his direct relationship with the Iowa State Highway Commission, he continued to devote his life to highway engineering and research. Marston retained his position as dean of engineering at Iowa State College and continued to present papers to the Highway Research Board. His research related to all facets of transportation, and he continued to speak on a regional and national level about transportation issues.[38]

Iowa engineers and the creation of the Highway Research Board

From the beginning, engineers at Iowa State College and the Iowa State Highway Commission conducted research that attracted national attention. For example, the highway commission's research on the cost benefits of dirt roads versus hard-surfaced roads received national coverage due to Fred White's role with the American Association of State Highway Officials (AASHO).

In 1924 White traveled to Washington, D.C., as a representative of AASHO to deliver testimony to the House of Representatives Committee on Roads. A *Service Bulletin* article summarized his testimony before the house and his research data on the value of surfacing roads. White's study argued that when traffic on a roadway reached 320 tons daily, paving the road became economical. Data collected by the highway commission and Iowa State College indicated that road improvements saved gasoline and tire wear.

In his testimony, White also asserted that highway work had become more efficient in the past 10 years, and he promoted

the work of the Bureau of Public Roads as an agent for highway research. White stressed the importance of coordinating research, referring to the independent but related nature of research in Iowa and Illinois. He expressed trust in the capacity of highway engineers to solve contemporary problems, such as the heavy loads of truck transport. White concluded his testimony optimistically: "We are studying these questions and we are confident that we are arriving at the solution of these various problems." He provided vision for the Iowa State Highway Commission in a new era of road surfacing, and Iowans led the drive to create a highway network.[39]

When White referred to coordinating highway research, he likely was thinking of the newly organized Highway Research Board. Today the annual meeting of the Transportation Research Board (the current name of the Highway Research Board) attracts an international audience of more than 6,000, but in 1924 the agency had served the highway engineering profession for a mere four years. The effort to create a national highway research agency, led by Anson Marston and the chief of the Bureau of Public Roads, Thomas H. MacDonald, had just come to fruition in 1920.

Marston's role in the creation of the Highway Research Board is most notable in his address to the annual meeting of AASHO in December 1919. The speech, delivered with the expectation of federal funding for highway construction, emphasized the need to coordinate highway research, investigation, and reporting. Such investigations were becoming increasingly important as states constructed paved and other hard-surfaced roads.

Marston advocated an agency to conduct "extensive scientific experimentation to develop the foundation principles of scientific highway engineering."[40] In his speech to AASHO, Marston announced a plan to generate highway research through eight entities. The research institutions and individuals would include AASHO, the Bureau of Public Roads, the engineering colleges and experiment stations, municipal testing laboratories, manufacturers' research departments, commercial laboratories, technical societies, and consulting highway engineers. Marston believed highway research by these entities should be directed by a central national agency to ensure that the work had a comprehensive scope and to avoid duplication.

By the time of his 1919 speech to AASHO, Marston was a nationally recognized leader in the highway research movement. Shortly after the speech he was appointed, along with H. H. Porter of New York, George Webster of Philadelphia, and Arthur Talbott of the University of Illinois, as a representative to the Division of Engineering of the National Research Council.

This group formulated a report to the chair of the division, suggesting that the National Research Council support and implement Marston's recently articulated plan for highway research, which included a recommendation to form six committees to oversee specific areas of research: economic theory, structural design, road materials, road construction, maintenance methods, and bridges and culverts. These specific topics encompassed all facets of highway engineering. The committees would facilitate systematic analysis of current research and help set the direction for the research needed to support future highway engineering.[41] Marston, Webster, and Talbott believed the National Research Council should direct the work, as the council had been created for just such a purpose.[42]

Upon examining the report, the National Research Council created a national Highway Research Committee, which met for the first time in November 1920 and which served as the foundation of the Highway Research Board (later the Transportation Research Board). Marston's efforts received additional recognition, as he chaired the Highway Research Board from November 1920 through 1923.

Marston also delivered papers to the board and served as a commentator on reports by other researchers. In particular, he further developed his theory of external loads on culverts and other closed conduits. This work culminated in a paper Marston delivered at the annual meeting of the Highway Research Board in 1929 entitled, "The Theory of External Loads on Closed Conduits in Light of the Latest Experiments." Developed from 21 years of research, the theories in this paper still serve as the model for theory on loads.[43]

Anson Marston, dean of engineering at Iowa State College (front left) Marston developed a theory of loads on culverts in the 1910s, and the Engineering Experiment Station provided the facilities to test concrete.

Thomas H. MacDonald was also a major promoter of the organization of the Highway Research Board. As chief of the Bureau of Public Roads in the U.S. Department of Agriculture, MacDonald proved integral to the board's development and success. The Bureau of Public Roads provided the majority of the funds for the board's work for several years. In 1922, for example, the Bureau of Public Roads contributed $12,000 of the board's $14,500 budget.

MacDonald also served as an ex officio member of the executive committee of the Highway Research Board until his retirement from the Bureau of Public Roads in 1953. Without the staunch support of MacDonald and the U.S. Department of Agriculture, the Highway Research Board would have faced great financial restrictions.[44]

Throughout the 1920s Iowa was a strong, if not dominant, presence on the Highway Research Board. Engineers from the Iowa State Highway Commission and Iowa State College served on Highway Research Board committees and presented papers at annual board meetings. The 1922 annual proceedings listed the members of the board and the committee members. Thomas R. Agg and Thomas H. MacDonald served on the executive committee, and Anson Marston represented the Association of Land Grant Colleges. Agg, Roy Crum, and Walter H. Root, maintenance engineer at the Iowa State Highway Commission, served on committees. Agg chaired the Committee on Economic Theory of Highway Improvement and served on the Committee on Highway Traffic Analysis. Crum assisted the Committee on Character and Use of Road Materials. Root chaired the Committee on Maintenance. The annual proceedings of the Highway Research Board reported formal and informal work conducted by Iowa State College engineers and researchers. One

report related preliminary work on soil investigation at Iowa State College. The board also appointed Iowa researchers to investigate specific questions.[45]

During the 1920s Iowa engineers continued to shape the Highway Research Board. Although the board promoted the dissemination of highway research among states, a coordinated effort had not been fully developed. To meet this need, MacDonald and Crum articulated a plan at the annual meeting of the board in December 1929. MacDonald prefaced Crum's paper with remarks emphasizing the value of a nationally coordinated program for highway research. MacDonald believed that much of the scholarship produced by highway researchers "repeated, revamped, and reworked" existing data. He believed that research should contribute new information to the profession, and the only means to produce such research would be a coordinated national effort as originally envisioned with the creation of the Highway Research Board.

MacDonald contended that research served as an investment, and the public would reward better service with increased confidence in highway engineers. He alluded to topics like soil engineering that he believed required attention and maintained that coordination of research was imperative. MacDonald concluded by stating that research often yielded unexpected results that were more important than the original intent of the investigation. Research and progress went hand in hand, he insisted, and with thorough research, engineers could determine highway policy.[46]

Crum followed MacDonald's introduction with the "Report of the Committee on Coordination and Program." In this report Crum outlined the activities of the Highway Research Board and suggested its future direction. These plans included a comprehensive program to compile references for research and a list of current projects, promote research by various agencies, conduct special investigations of national scope, study the work of various research agencies, and disseminate research information. He then detailed six primary fields for investigation: highway administration and finance, highway transportation, highway design, materials and construction, maintenance, and traffic. Each topic had subordinate areas for investigation and research; for example, highway design listed research subfields such as roadway, drainage structures, and surfaces. Crum's paper established a detailed national agenda for highway research.[47]

Machinery and maintenance

Locally, Iowa highway engineers focused their attention on highway safety. In 1925 Iowa adapted machinery and equipment to mark center lines on highways. The Iowa State Highway Commission modified a World War I surplus truck to apply center lines on paved primary roads.

The marking crew consisted of a driver and a person at the painting wheel, mounted just ahead of and outside the rear wheel on the driver's side of the truck. The truck had an adjustable arm with a wheel on the passenger's side to act as a guide outside the curb, ensuring the line would always be painted at equal distance from the curb. The truck traveled at a velocity of two miles per hour, but occasionally on curves the workers had to manually mark their path, slowing their progress. The Iowa State Highway Commission devoted three months in the 1925 and 1926 painting seasons to mark 1,100 miles of pavement. This unique vehicle demonstrated the creativity and ingenuity of highway commission workers in serving Iowans, traits that continued throughout the existence of commission.[48]

Slope cutter, Monona County, Iowa, c. 1925
In western Iowa the rolling loess hills challenged contractors to develop economical methods of cutting through the hillsides. This machine saved funds and labor compared to cutting the hillside by hand.

Pavement settlement proved to be a universal problem common with the new portland cement concrete roads. Before 1930 maintenance engineers and workers developed three techniques to raise settled pavement: filling the area with a bituminous (asphaltic) mixture; raising the pavement with jacks and filling the area underneath with earth or sand forced in by compressed air; or breaking up the old slab, filling in the subgrade, and constructing a new slab. These methods proved ineffective, inconvenient, or cost prohibitive.

In the late 1920s Iowa State Highway Commission mechanic John Poulter began to experiment with a device to raise pavements through hydraulic pressure. His first machine consisted of a tractor valve and valve guide, which he

Iowa State Highway Commission concrete slab tests, c. 1930
The creation of the national Highway Research Board increased the importance of systematic highway research. To test concrete, the Iowa State Highway Commission laid sample sections of pavement to determine the durability of concrete to provide the greatest economy from Iowa highways. Highway commission workers used World War I surplus trucks to haul the concrete slabs to the test labs.

27

Center line marking, c. 1927
During the summer highway commission crews marked center lines on paved primary roads. The truck had an arm with a painting wheel and traveled at about two miles per hour.

Mud pump, c. 1931
John Poulter, an Iowa State Highway Commission employee, developed a machine called the mud pump or mud jack to raise sunken pavement. The pump forced a slurry through pre-drilled holes in the pavement, lifting the pavement.

grouted into a pre-drilled hole in the slab. After preliminary success, Poulter applied for and received a patent for the pump and implemented the principle on a larger scale. Poulter's two-cylinder pump received its power from a 20-horsepower gasoline engine and consisted of the following parts: a hopper to receive the earth, water, and cement; a mixing chamber; a receiving chamber for the mixed materials; the pump; an outlet hose; and the power supply. [49]

Walter H. Root, Iowa State Highway Commission maintenance engineer, reported the initial use of Poulter's "mud pump" at the December 1930 meeting of the Highway Research Board. Root summarized the process: To raise the pavement, workers cut a four-inch expansion joint across the pavement, drilled two-and-a-half-inch holes in the slab, and placed a hose and nozzle in the holes. After a worker had thus prepared the section of sunken road, an operator activated the pump. The workers filled the hopper with earth, water, and port-land cement at a ratio of one part cement to 20 of earth and water, and the pump forced the slurry through the hose and under the pavement. The slurry provided resistance, raising the pavement.

Root commented that black topsoil and loess proved the most satisfactory type of earth content. Sand wore out the cylinders of the pump, gravelly soil clogged the valves, and clays did not produce the necessary "creamy grout." Early trials determined that without the cement mix the mud slurry would escape from holes or from under the edge of a slab. If the maintenance crew waited two hours, the slurry achieved a sufficient set so that pumping could be resumed. The process of mud pumping proved effective and, because it allowed traffic to continue on the other lane of pavement during the repair process, relatively convenient.

Root's paper received attention in high-way engineering literature. A report in *Roads and Streets* in April 1931 related that Poulter had associated himself with the National Equipment Company of Milwaukee, Wisconsin, to produce the mud pump commercially. The previous issue of *Roads and Streets* contained a listing in the "New Equipment and Materials" section for the National Equipment Company's "mud-jack." The company continued to develop versions of the mud pump, and a smaller version, which raised curb, gutter, and sidewalk sections, came on the market in 1934.[50]

In an essay in *American Highways*, published by AASHO, Root described Iowa's maintenance program for the year 1930. The Iowa State Highway Commission used the mud pump on 200 settlements, raising 9,292 linear feet of pavement as much as 13 inches. The commission calculated that the process cost $1.02 to raise a square yard of pavement, slightly more than a blacktop pavement fill, but the pavement stabilized by the mud pump proved less liable to re-settle. Root concluded his report by stating that "such procedures would result in greater safety to the traveling public and in greater economy to the State."[51]

Controversy and the Iowa State Highway Commission

In 1933 the highway commission and Chief Engineer Fred White faced an assault by the Iowa legislature when the Iowa house accused White of colluding with portland cement concrete manufacturers. Outraged at the charges, White demanded a full hearing in the house.

The accusations followed a report in Missouri citing evidence of inside deals between the Missouri State Highway Commission and concrete supply companies. The *Des Moines Register and Tribune* led an investigation into the Iowa situation to determine whether a similar conspiracy existed. A house committee determined that evidence suggested collusion and neglect and that White had been at fault for not notifying the governor or attorney general of the nearly identical costs in bids from contractors.

During the controversy many newspapers printed editorials portraying White as a promoter of the public welfare, and most of them advised readers to be cautious and let the facts be presented. But a few (primarily newer) newspapers suggested that initial reports indicated collusion, and one newspaper implied that corruption had crept into other highway

commission practices as well. The *Dyersville Commercial* charged that the highway commission regularly overspent funds on projects and that during the 1920s this practice had wasted the county's money.

After White himself addressed the Iowa house, documenting the commission's contracting processes and requesting a

Iowa State Highway Commission workers prepared pavement for raising, c. 1931
Maintenance workers drilled holes in the pavement about three feet apart to inject the slurry.

Iowa State Highway Commission workers raising pavement with a mud jack, c. 1931
The mud jack worked even while traffic continued on an adjacent section of pavement.

Longitudinal Section Through Dip

Drilled Holes Through Which Mud Is Pumped

℄ Road

4" Cut Expansion

Longitudinal Section and Plan Showing Preparation for Raising Slab

Longitudinal Section After Slab Has Been Raised

29

Walter H. Root
Root served as the first maintenance engineer for the Iowa State Highway Commission and was the chairman of the Highway Research Board in 1954 until his death the same year.

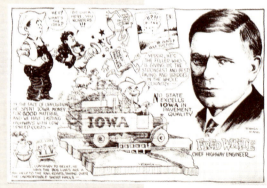

Fred White cartoon, c. 1935
White, chief engineer of the Iowa State Highway Commission, often received favorable press as evidenced in this cartoon. The work of the commission, however, did not always receive complete public support, and White occasionally defended the agency before adversarial legislators.

full hearing, the Iowa house cleared him of all charges and ordered them stricken from the record.[52] White's confrontations with the legislature were not to end soon, however, and documented research proved invaluable when defending the commission's materials purchases or, as in the following case, the purchase of a gravel pit.

In the fall of 1933 White returned to the state capitol in Des Moines to defend the purchase of a gravel pit in Guthrie County. Western Iowa had few sources of aggregate and, in 1928, the highway commission had purchased land that it claimed had quality ledges of gravel for concrete road construction. Representative Aldrich purported that the state had wasted money on this purchase because the gravel did not prove suitable for concrete.

White skillfully testified at the hearings and used research to support the purchase. Badgered repeatedly to admit how the state had been boondoggled, White pointed out that most gravel in its natural state is unsuitable as an aggregate; it must be washed to remove shale and other impurities and screened to separate the aggregate into the proper size and proportion. In his testimony, White explained that the state had removed numerous samples of gravel from the Guthrie County pit, screened and washed it, and conducted freeze-thaw tests. He added that crushing tests of concrete made from

the samples showed the material satisfied the requirements as a durable aggregate. The chief engineer acknowledged that the pit had not yet produced any usable gravel, but the highway commission's purchase of this pit had caused Des Moines contractors to lower their prices. White's integrity and expertise—and well documented research—protected the reputation of the highway commission and demonstrated leadership in the face of what seemed like a witch-hunting legislature.[53]

Toward better roads

The highway commission's annual report of 1929 noted that for the first time Iowans had "adopted for themselves a policy of constructing a system of paved roads." A state primary road bond issue had passed in November 1928 to provide $100 million for road construction. The state supreme court had ruled the bond act unconstitutional and invalid, but the 1929 Iowa legislature passed legislation permitting the issuance of county primary road bonds. After this act, all but one of the Iowa counties authorized a total of more than $118 million in county primary road bonds. Through these monies, contractors completed almost 740 miles of concrete paving in 1929, 700 in 1930, and 1,000 in 1931, and Iowa's reputation as a concrete state was established for the next 20 years.[54]

Collaborating with private industry

Besides paving and graveling primary roads, the commission had thousands of miles of secondary roads to maintain. In an effort to demonstrate the most modern methods of earthen road construction, the commission cooperated with private industry. Two firms, the Adams Blade Grader Company and Holt Manufacturing Company, provided equipment to build a three-mile stretch of test road south of Ames. The companies attempted to meet commission standards and document expenses, and they provided all work free of charge. This road, the commission said, represented conditions across the state reasonably well. The commission provided someone to observe and record data related to the work.

The trial proved useful in determining cost factors, especially that time delays while waiting for materials or weather nearly doubled the cost of the work. The commission further concluded that the blade grader provided the greatest cost benefit for the work accomplished. Through collaboration with private enterprise, the commission demonstrated useful methods of local road maintenance.[55]

Batching

Iowa's policy of building hard-surfaced roads of portland cement concrete necessitated a reliable method for selecting adequate aggregate. A development promoted by the Iowa State Highway Commission in 1927 allowed contractors to portion gravel by weight, known as "batching," instead of by volume. Batching aggregate by weight provided a more uniform concrete, especially if the contractor sometimes removed aggregate from the pit and let it sit—and therefore settle—in a rail car overnight.

Fred White reported this improvement in an essay in the road periodical *Roads and Streets*. White concluded by commenting on the continuity of commissioners in Iowa and the excellent reputation the Iowa State Highway Commission had in the counties. White boasted of the 100 percent rating the test lab had received from the Bureau of Public Roads. He demonstrated that, through proper administration and controls, Iowa engineers contributed methods and models useful for other states.[56]

The Department of Materials and Tests

In addition to the *Service Bulletin*, beginning in 1920 the Iowa State Highway Commission printed an in-house *Monthly Newsletter*. This newsletter served as a vehicle to disseminate commission information and the occasional humorous anecdote. The newsletter reported the status of projects like field surveys and the creation of a map of the primary road system, and it provided employees with information about the work of new departments.

Blade grading a roadway, c. 1925
The Iowa State Highway Commission hired contractors to create and maintain gravel and earthen roads. The commission determined that the blade grader provided the greatest cost benefit for maintenance.

One new department was the Department of Materials and Tests, established in 1919 and directed by Roy Crum, who had left the Engineering Experiment Station at Iowa State College to lead materials research at the highway commission. Now at the commission, Crum distinguished himself as an important highway engineer. Under his guidance as its first

Aggregate at rail yard, c. 1925
In areas without a nearby supply of aggregate, material was hauled by rail and dumped at work sites. A crane, such as this one in northwest Iowa, dumped the aggregate into a large hopper, which in turn portioned the aggregate into dump trucks for the haul to the roadway.

Iowa State Highway Commission test laboratory, c. 1925
In 1923 the highway commission moved to its own facilities on Lincoln Way in Ames. The commission now had its own research facilities but maintained its close working relationship with Iowa State College.

chief, the Department of Materials and Tests continued and broadened the highway commission's focus on investigating road building materials begun when the commission was part of Iowa State College.

The Department of Materials and Tests systematically investigated, tested, and recorded the supply and quality of materials for hard-surfaced roads. On November 15, 1919, the department assumed responsibility for testing materials used in state-funded road projects and sometimes those used in county projects. Prior to this date, the bulk of material tests had been conducted by the Iowa State College Engineering Experiment Station under Anson Marston's direction. The creation of the department further established the commission as an independent agency and reporter of information. The Department of Materials and Tests would provide enduring contributions to road building on the state and national levels.

Aggregates

Because federal aid and Iowa's Primary Road Act funded only hard-surfaced roads, the Department of Materials and Tests became increasingly important to identify appropriate materials for building hard-surfaced roads. Detailed tests provided information to engineers who supervised construction, and retention of these data proved valuable for the long and short terms. As the highway commission

Roy Crum conducting field investigation, 1921
Roy Crum (right), the Iowa State Highway Commission's engineer for materials and tests, later became the director of the national Highway Research Board. He developed a method for portioning aggregate by weight rather than by a coarse gravel-to-sand ration. The method became a standard practice.

oversaw the construction of the first generation of concrete roads, engineers and scientists used the tests to determine the location of suitable aggregates on a local level. Later, the tests proved useful in selecting ledges of limestone that would provide the most durable aggregate and create long-lasting pavements.

Immediately after departing from the Engineering Experiment Station, Crum authored an important station bulletin that established methods for determining the volume of aggregates used in concrete based on Iowa aggregates with a higher percentage of sand. Crum explained that the two necessary components of good concrete are quality materials and proper manufacturing conditions. He determined that gravels could be graded or classified as either sand (called fines or fine aggregates) and pebbles (called coarse aggregate). Previous tests had sifted the gravel into numerous components, but this simpler classification method proved equally successful. Crum's tests used data gathered in road construction and demonstrated the results applicable for practical use.[57]

As chief of the Department of Materials and Tests, Crum provided leadership to longtime commission employees Bert Myers and Mark Morris. Myers and Morris served long careers at the commission and contributed work of local and national significance. When Crum left the highway commission to direct the Highway Research Board in 1928, Myers became chief engineer for materials and tests.[58]

Concrete mixes

Bert Myers directed the Department of Materials and Tests from 1928 to 1959. Myers grew up in Dallas County and attended Iowa State College, where he received a degree in 1917. He served a brief tenure as assistant county engineer in Maquoketa, Jackson County, Iowa, before joining the highway commission in 1917. He worked at the commission until his death in 1959. Myers served as a principal reporter for the Iowa State Highway Commission at the Highway Research Board, participating in discussions and delivering papers to the board. Specializing in concrete mixes and materials for pavements, he served on Highway Research Board committees related to those topics, continuing the Iowa State Highway Commission's strong presence on the board's national committees.[59]

Traffic counts

Equal to the tenure of Bert Myers was that of Mark Morris. Morris grew up in southeast Iowa near the town of Stockport. He served in various engineering positions until permanently joining the Iowa State Highway Commission in 1919. Morris had worked for the highway commission in the summers of 1914 and 1915 while attending Iowa State College but soon withdrew from the college and left the commission. He rejoined the commission in 1919 and finished his work at the college, graduating in 1921. After graduation, he attained a position as research assistant to the engineer of materials and tests. Except for a short leave from July to December 1925 to investigate culvert work for the Highway Research Board, he stayed in Ames until his retirement.

Like Bert Myers, Morris also represented the Iowa State Highway Commission at Highway Research Board meetings and frequently presented papers. He first represented the commission in 1925 with a paper titled "Progress Report on Culvert Investigation," a study completed with Roy Crum, and gained a respected reputation in the field of research and investigation. He served on Highway Research Board committees, including the Culvert Rating Committee, the Concrete Curing Committee, and the Highway

Capacity Committee, and also received appointments to the National Research Council. On a local level, he worked to improve the quality of life in Iowa by advising the Iowa Emergency Relief Administration on highway matters and working with Iowa State College students as a contact person for the American Society of Civil Engineers.[60]

Morris's role with the relief administration concerned traffic studies on Iowa highways. The Iowa State Highway Commission's first 24-hour traffic count project was in 1934. The commission paid for equipment, and the federal and Iowa

Bert Myers
In 1928 Myers succeeded Roy Crum as engineer for materials and tests at the Iowa State Highway Commission. Myers directed the department until he died in 1959.

emergency relief administrations paid for the staff to conduct the count. The commission established one master station south of Ames, which counted traffic 24 hours a day for a year. The commission also established 151 additional 24-hour stations, but these posts operated only 14 days. Another type of station, the 12-hour station, counted traffic for six days, and the commission conducted 1,178 12-hour surveys. The Iowa State Highway Commission also employed counters at 81 miscellaneous stations. This work served as a blueprint for other states monitoring road traffic.[61]

The road station south of Ames provided the first 24-hour traffic counts for a study to determine the traffic volume on the primary road system. The study proved useful in determining the true amount of traffic on a primary road. The Iowa State Highway Commission correlated the data gathered at this master station with those of the other stations. The primary findings showed that almost three-fourths of the traffic consisted of passenger vehicles. The report also revealed the local nature of traffic: 33 percent from the county and 87 percent from inside the state of Iowa.

This research enabled the commission to determine road use and develop more accurate methods for addressing highway needs and finance, and it showed that data gathered at any station could represent general traffic patterns; agencies need not maintain a master station to develop reliable statistics on traffic volume.

Although the commission admitted the sketchy nature of information from outlying stations, the Iowa State Highway Commission tabulated the data, provided useful conclusions, and planned for additional research with the Bureau of Public Roads.[62]

A comprehensive textbook

Even though the Iowa State Highway Commission now had its own research department, the Iowa State College Engineering Experiment Station continued to conduct investigations for the commission. The Good Roads section of the experiment station, for example, investigated mundane subjects like bridge paints. Although the state had established paint standards by 1913, this research provided additional information on the durability and protective nature of the paints. The experiment station also began to direct origin and destination studies and weight studies.

In 1919 Thomas R. Agg concluded in a research paper that actual traffic weights on roads exceeded conventional estimates. Agg foresaw that truck traffic, as well as truck size and weight, would increase dramatically as highway surfaces and durability improved. His predictions fostered the study of the load-bearing capacity of pavements.

Agg also expressed frustration with trying to maintain gravel roads for motor vehicles. His research showed that most traffic on Iowa roads originated in Iowa and that better highways, rather than serving primarily tourists as some people argued, benefited all the people of Iowa.[63]

Agg had joined the Iowa State College staff in civil engineering in 1913, but his ties to the college went back to his undergraduate training. After receiving his degree in 1905, he joined the faculty of the University of Illinois. In Champaign, Agg taught engineering drawing and worked for the Illinois State Highway Commission during the summer months. In 1909 Agg accepted a permanent position with the Illinois State Highway Commission to direct experimental road construction, a job he held until he returned to Iowa State College.

Agg's leadership in scholarship, education, and research positioned Iowa State College and the Iowa State Highway Commission at the forefront of highway engineering. He joined the Engineering Experiment Station staff in 1913 as road engineer and that year published his first research bulletin, an essay on Iowa gravels for concrete. He served on the Iowa State College faculty, where he succeeded Marston as dean in 1932, and remained on the faculty until 1946.[64]

In addition to his research and teaching, another of Agg's significant contributions to highway development was a standard text on highways, *Construction of Roads and Pavements,* which he authored in 1916 and which remained in print through 1940. This comprehensive book covered all elements of highway engineering, focusing especially on various surfaces and providing a concise description of portland cement concrete road construction.

According to the text, concrete could be placed through either a dry-batch or a wet-batch method. In the dry-batch method, material was taken to the road site and there water was added to it. In the wet-batch method, road builders mixed the aggregate and water at the railroad siding or gravel pit. The dry-batch method proved useful for rural highway construction, particularly when water was nearby. Contractors used wet-batch construction in cities where the haul was short. After the builders placed the concrete, the next step was to finish the pavement with a tamping machine

that compacted and smoothed the slab. Finishing machines worked on pavements up to 24 feet wide, so occasionally the concrete had to be tamped by hand. The quality of hand finishing depended on the skill of the worker. These methods of concrete construction described in Agg's text changed little through the 1930s.[65]

Agg contributed to highway research on a national level through the national Highway Research Board. He held positions on the executive committee until 1943, a tenure difficult to match, and chaired the Committee on Economic Theory and the Committee on Highway Transportation Costs. He also provided commentary on various papers presented at the Highway Research Board's annual meeting. His work for the board received the recognition of his peers in 1936 when he received the George Bartlett Award, the second Iowa State College graduate to do so in the six-year history of the award.[66]

Pavement life

While the Iowa State Highway Commission studied traffic volume, the Engineering Experiment Station documented the service life of pavement. Iowa State College civil engineer Robley Winfrey provided direction in the research of pavement life and transportation costs. Winfrey grew up in Hastie and Des Moines, Iowa. He finished coursework in civil engineering in 1918 at Iowa State College and worked briefly for the city of Des Moines. When Winfrey returned to Ames, he joined the staff of the Engineering Experiment Station as assistant bulletin editor. Evidently

Thomas Redford Agg
Thomas Agg succeeded Anson Marston as dean of engineering at Iowa State College. Agg, as well as Marston and Thomas MacDonald, established the Highway Research Board as a major body.

35

Winfrey did not find himself restricted to office work, because he helped Thomas Agg on tractive resistance and fuel consumption studies.[67]

Through the 1920s Winfrey devoted most of his time to editing Engineering Experiment Station bulletins and answering correspondence. Gradually he received more responsibilities until, by the mid-1930s, he was the station administrator. In 1935 he took a leave of absence to coordinate traffic studies in 45 states for the Bureau of Public Roads to determine the economics of transportation on a national scale. Winfrey took particular interest in road life studies and presented papers at the Highway Research Board meetings of 1935 and 1940.[68]

The initial report of the 1935 Bureau of Public Roads study led by Winfrey included data from New York, Iowa, Michigan, Massachusetts, and Rhode Island. The test relied on five methods for determining average road life: individual unit method, original group method, composite original group method, multiple original group method, and the annual rate method. Although any of the methods could be used to determine pavement life, Winfrey suggested that the individual unit method proved least accurate. Winfrey included reasons for retirement (the end of a road surface's usefulness) in the study. The causes of retirement included surface and structural failures, obsolescence, highway improvements, non-highway construction or catastrophe, and sale or transfer of a roadway to another governing authority. Winfrey studied the life of actual pavements so that real-world situations, rather than artificial laboratory tests, would be the basis for data. Winfrey developed an informational base on pavement life and continued to collect more data.[69]

In 1940 Winfrey provided specific summary data for road surface life at the Highway Research Board's annual meeting. Presented with Fred Farrell, an associate highway engineer with the Bureau of Public Roads, this report established general limits for the life of surfaces on primary rural highways. The survey revealed the life expectancies of various roads: earthen, gravel, bituminous surface treated, mixed bituminous, bituminous penetration, bituminous concrete, portland cement concrete, and brick. His data faced revision but served as the first comprehensive study of pavement life. Iowa engineers continued to develop the necessary research for an economical and durable highway system.[70]

Highway surfaces

While Winfrey focused on pavement life, Professor Ralph Moyer studied road surfaces for roughness and skidding characteristics. Moyer received his bachelor's degree from Lafayette College in Easton, Pennsylvania, and began his tenure at Iowa State College in 1921 as an instructor in the civil engineering department. He received a master of science degree from Iowa State College in 1925 and a professional degree in civil engineering in 1934. Moyer worked for the Engineering Experiment Station, concentrating on skid testing for various road surfaces.

When he delivered his first paper to the annual meeting of the Highway Research Board in 1933, Moyer became an expert in skid testing and research. He continued to research and present papers related to road surfaces, traffic safety, and winter maintenance, among other topics, attaining such a reputation that he received an appointment as the first director of the Institute of Transportation and Traffic Engineering at the University of California-Berkeley in 1948. Iowa State College-trained engineers continued to attract the attention of highway engineers across the nation.[71]

Soils engineering

Merlin Spangler, a highway commission employee from 1919 to 1924 and later an Iowa State College civil engineer for the Engineering Experiment Station, grew up in Des Moines and received his civil engineering degree from Iowa State College in 1919. After a five-year stint with the Iowa State Highway Commission working on bridge design, he returned to the college, received a master's degree in 1928, and finished his career at the college.

As assistant engineer at the Engineering Experiment Station, Spangler performed early work on the subject of pavement loads, focusing on stresses on concrete pavement, and later became a noted soils engineer. In 1935 at the annual Highway Research Board meeting he presented a paper discussing the reactions of the subgrade beneath concrete slabs to various stresses. He asserted that conventional scholarship had yet to determine the varying stresses tolerated by pavements. Spangler established additional results for drainage pipes.[72]

Summary

By the eve of World War II, Iowa highway engineers had firmly established their leadership in highway research and innovation. Anson Marston and Thomas H. MacDonald led their national counterparts in the creation of the Highway Research Board, and an Iowa State College graduate, Roy Crum, became the board's director in 1928. Crum provided leadership for the Highway Research Board for 22 years.

The integrity and vision of these leaders and many others were recognized by their peers, and three Iowa State College graduates would earn the George Bartlett Award, the Highway Research Board's award for outstanding contributions to highway progress, during the first eight years the award was presented. The Iowa State Highway Commission developed a tradition of experimentation with equipment and demonstrated that Iowa engineers looked beyond the present, seeking improvements for the public and industry. Engineers like John Poulter developed machines like the mud pump, and private industry adapted the machinery for commercial use. Iowa engineers proved they could hard-surface the state road system and lead the nation in national highway planning. These advances foreshadowed additional achievements through research in the coming decades.

Such innovation was to lead to Iowa's greatest contribution to portland cement concrete road construction: the slip-form paver.

Chapter 3

Improving the Road System: 1939–1950

The year 1939 inaugurated another period of road designation and construction for the Iowa State Highway Commission when farm-to-market roads were established as a county system to truly connect farms and communities. By 1950 Iowa State Highway Commission engineers had developed a functional slip-form paver that served as a prototype for modern machines. Iowa State College engineers continued to investigate such topics as soil engineering, traffic safety, and cost accounting and, as traffic demands increased, Iowa engineers continued their diligent public service, promoting economical and safe highway construction.

In 1940 the Highway Research Board established an award to recognize research of outstanding merit. Three of the first six recipients were Iowa State College engineers: Merlin Spangler, M. B. Russell, and Ralph Moyer. The state would establish its own Iowa Highway Research Board to meet the challenges related to the increased use of Iowa's highways.

The farm-to-market system

The first farm-to-market bill passed the Iowa legislature in 1939 to complement federal aid appropriated for secondary roads. Typical of all federal support of the day, the federal dollars were to be matched by state funds. The act included three provisions: it designated a farm-to-market road system composed of 10 percent of the state's secondary roads, established a farm-to-market road fund, and provided for maintenance and improvement of the farm-to-market roads. The law also limited the appropriations devoted to the primary road fund. From 1940 through 1949, the primary road fund could not exceed $17 million; any surplus had to be spent on secondary roads. The state anticipated an allocation of $1,679,358 for the period 1938 to 1940. In actuality, the state received more than $2 million in federal funds in the three-year period.[73]

Due to the Farm-to-Market Act of 1939 and its focus on secondary roads, construction and maintenance on primary roads suffered. The Iowa State Highway Commission opposed this redirecting of funds from the primary road fund to the farm-to-market program but failed to stop it because the legislators promoting the secondary road system believed that the primary road system had nearly been completed. This proved untrue, and in 1947 the Highway Investigation Committee recommended an appropriation of more than $480 million for the next 20 years for primary road construction.

In the revised farm-to-market legislation of 1947, the state established new guidelines and funding, requiring that the funds be allocated and the mileage designated equally among counties. By 1949 the primary road fund regained lost revenue. A road use tax fund was established by the legislature to provide additional revenue, and the $17 million cap for the primary road fund was removed. The road use tax fund included registration fees, a four-cent fuel tax, and

Rainbow arch bridge, c. 1925 (facing page)
Rainbow arch bridges were so named for their unique shape.
This one, in Marshalltown, Iowa, has been demolished.

10 percent of the funds from the state sales tax. These new resources, in addition to a $5 million allocation from the legislature, increased the primary road fund to more than $10 million.[74]

Modifying equipment to serve special purposes

Iowa State Highway Commission road engineers were an innovative lot, refusing to settle for inadequate or expensive construction methods. One small example of their creativity and persistence is their solution to curb removal.

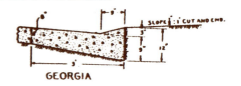

GEORGIA

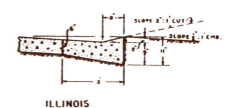

ILLINOIS

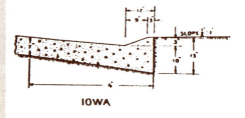

IOWA

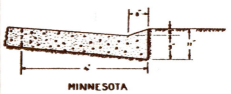

MINNESOTA

From the beginning of portland cement concrete road construction, Iowa's rolling hills created a drainage problem for highway engineers. A common solution to drainage problems in a number of states, including Illinois, Minnesota, and Missouri, was the integral curb, also called a lip curb. Lip curb construction was first used in Iowa in 1920 in the western loess hill region to remove water from pavement without eroding the shoulders or the road. The practice spread throughout the state, and in 1948 Chief Engineer Fred White estimated that 40 percent of Iowa's highway pavements—roughly 2,200 miles—had lip curbs.

Lip curbs, c. 1930
Integral curbs, called lip curbs, were added to many Iowa pavements in the 1920s and 1930s to drain highways and prevent shoulder erosion. The curbs adequately drained the pavements but were a hazard to motorists.

While solving one problem, however, integral curbs created numerous other difficulties, particularly for safety. If a vehicle rode up on the curb, maintaining control was difficult. As vehicles got larger, curbs made relatively narrow roads seem even narrower. From 1920 to 1940, the number of motor vehicles on Iowa roads and their average annual mileage doubled. The average size and weight of motor vehicles increased, and the traffic volume of larger vehicles grew by more than 30 percent. Iowa's narrow, curbed roads were taking a beating. To quote engineer White, "Time, the elements, and the pounding of traffic have taken their toll. These old pavements must be widened, resurfaced, strengthened, and modernized." Integral curbs had become obsolete.

Iowa engineers investigated other states' methods of lip curb removal and found that the use of the air hammer predominated. Although air hammers removed lip curbs effectively, the process was time consuming and costly. If the highway commission wanted to improve its existing road system, a faster method had to be developed.

In January 1948 the Iowa State Highway Commission acquired a pavement breaker manufactured by the R.P.B. Corporation of Los Angeles. The machine smashed pavement with a steel piston inside a cylinder six inches in diameter and 48 inches long. The steel piston head weighed 260 pounds and received different tools depending on the nature of the work. Compressed air caused the head to strike a quick sharp blow, the "equivalent of Paul Bunyan swinging a two-hundred and sixty pound sledge."[75] Iowa State Highway Commission officials decided to modify the machine to hit a blow horizontally and knock the lip curb off of pavement. Charles L. Gleason, the commission district engineer for Ames, received the assignment and designed a mount to attach the breaker to the rear of a truck. Charles Kinderman and Joseph Gibson of the commission machine shop constructed the mount and put it in working order. These Iowa innovators, like so many before and after them, did not accept conventional practice and developed a creative solution to a specific problem.[76]

A pivotal invention: the slip-form paver

Innovations like the modified pavement breaker were common, so when Iowa engineers tested the slip-form paver in 1949 it was not at first recognized as a pivotal contribution to highway construction. Developed entirely at the Iowa State Highway Commission laboratory in Ames, the slip-form paver allowed road builders to construct portland cement concrete roads without using fixed forms. This allowed contractors to significantly increase the amount of pavement laid in a day.

The introduction of the slip-form paver created little stir initially. The October 5, 1949, headline of the *O'Brien County Bell* shouted "SPOIL MILFORD HOMECOMING 33-19"; an accompanying story described the local high school football team's victory in detail. Below the fold ran a short article: "County Engineer Explains Paving Project Here with New Machine." This "new machine," arguably Iowa's greatest contribution to highway construction, was to alter the nature of portland cement concrete road construction in its entirety.[77] On a good day in 1949, a crew could lay about 1,000 feet of concrete using fixed forms. Today's modern slip-form paver can lay a slab of concrete 12 inches thick and 12 to 18 feet wide at the rate of a mile or more a day.

The development of the slip-form paver began in 1946 when the laboratory chief at the highway commission, James Johnson, and two other commission

Lip curb removal, c. 1950
In 1949 ISHC Engineer C. L. Gleason and other ISHC employees modified a pavement breaking machine to rapidly remove the lip curbs. The curb remover allowed the ISHC to widen miles of Iowa pavement.

employees, Rudy Schroeder and Willis Elbert, attended a demonstration of cement-treated base construction. After witnessing the demonstration, Johnson suspected that a mix with an increased proportion of cement that was vibrated into place by a machine would eliminate the need for fixed forms. In 1947 these engineers experimented with their idea and constructed a small prototype that extruded a slab of concrete 18 inches wide and three inches deep. The experiments continued, and the men built two larger models. The second model increased the pavement width to three feet, and the final machine paved a 10-foot section six inches deep.[78]

The engineers attached the 10-foot paver to a Ford V-8 engine, a Ford transmission, and a gear box from a concrete saw. The gear box drove a set of dual pneumatic tires in the rear; a set of single truck wheels in front completed the machine. The team fitted the machine with a second engine to vibrate the concrete. The commission tested the paver three times and found its performance satisfactory for use on a public project, a stretch of road in Primghar in northwest Iowa's O'Brien County.[79]

First prototype slip-form paver, c. 1947
Slip-form paving, a process that revolutionized portland concrete cement paving, was developed by Iowa State Highway Commission engineers. The first model of the paver laid an 18-inch wide, three-inch deep strip of pavement. Final improvements allowed contractors to pave a mile of roadway in a day.

Prototype slip-form paver, c. 1948
The second model of the slip-form paver laid a section of pavement 36 inches wide.

In September 1949 the one contract bid received by the highway commission to pave a half-mile section of highway through Primghar was rejected based on cost. As a result, the commission, the O'Brien County board of supervisors, and Primghar officials decided to experiment with the new slip-form paver. The commission had little time to complete the paving project because it had committed

the machine to lay concrete in Cerro Gordo County in the central part of the state on October 9. Grading of the Primghar road began on September 19; paving began on September 28 and was finished on October 1, well in advance of the Cerro Gordo County project.

Its speed notwithstanding, this first slip-form paving project did not proceed without complications. Because the paver produced a section 10 feet wide, a single lane was created by laying two sections side by side, leaving a three- to four-inch gap between the sections that workers had to fill later. Hairline cracks developed in the surface of the pavement, and engineers worked to diminish the level of cracking.[80]

The commission used this slip-form paver only three more times. In October 1949, as promised, the machine was used in Cerro Gordo County to lay one mile of pavement. Four years later the Iowa State Highway Commission laid a concrete base with the paver in eastern Iowa on Highway 30 in Cedar County. In 1954 the commission leased the machine to Ray Andrews, a private contractor, to pave a road in Greene County. The contractor altered the machine, removing the hopper, and the concrete was dumped directly on the grade much as it is today.

By 1955 commercial firms had developed functional slip-form pavers. The Quad Cities Construction Company produced a paver that laid two full lanes of pavement and advanced on crawler tracks rather than wheels. The Quad Cities Construction Company completed approximately 28 miles of slip-form paving in Iowa in 1955. That year, highway construction crews used a slip-form paver in Colorado and Wyoming, and a commercial need developed for this technology.[81]

Thomas H. MacDonald and federal highways

As chief of the Bureau of Public Roads, Thomas H. MacDonald led the national effort to create federal funding for interstate highways. In the 1940s engineers witnessed growing tension between rural and urban factions over road-building funds, and in 1944 MacDonald attempted to build a consensus between these groups. As congress debated funding formulas, road builders and other concerned parties like Fred Brenckman of the national Grange supported MacDonald's efforts. But MacDonald's plan to appease both the rural and urban constituencies was not adopted by congress.[82]

During his long career, MacDonald had developed significant credibility and was often in the national spotlight. The *Saturday Evening Post* ran a feature on MacDonald in 1944 depicting him as an omniscient engineer on whom the public could rely for solutions to modern road planning and construction questions.[83]

The story focused on MacDonald's role as chair of the Interregional Highway Committee. President Franklin Roosevelt had appointed the committee in April 1941, and it released its report on a national highway system to the president in January 1944. As chair of the committee, MacDonald argued for roads that met traffic needs and were financed by road user taxes on such items as fuel and tires. The *Post* story mentioned that the Bureau of Public Roads and the Interregional Highway Committee had developed a national plan without bias toward states that already had substantial transportation networks in place; all states received equal consideration in the planning.

Slip-form paver, c. 1949
The final Iowa State Highway Commission slip-form paver could pave one lane of traffic in a single pass. The machine paved a 10-foot section and returned to pave the adjacent lane. Workers then filled the gap between the two strips.

Toward better roads

During the 1940s Iowa researchers and engineers continued to participate in the national Highway Research Board meetings in Washington, D.C. Iowa State College professors Merlin Spangler, Robley Winfrey, and Ralph Moyer also maintained a presence in research through the 1940s. In 1943 Spangler and M. B. Russell's paper on soil moisture received the Highway Research Board award for research of outstanding merit. Moyer continued to demonstrate his knowledge of road surfaces and safety matters, presenting papers in 1940, 1942, 1945, and 1947. When he did not present, Moyer often served as a commentator for other papers. In 1943 he received the Highway Research Board Award for his 1942 paper on motor vehicle operating costs. He joined the Highway Research Board executive committee in 1945 and received a second appointment in 1947.[84]

Iowa's prominent presence at the national Highway Research Board was fueled by the efforts of researchers and engineers at Iowa State College and the Iowa State Highway Commission to continue to improve methods and materials for building better roads in Iowa. Their

Merlin Spangler, c. 1970
Spangler, Iowa State
University engineering
professor, investigated soils
engineering and in 1951
authored the definitive
work Soil Engineering,
still in print.

research focused on soils and road bases, portland cement and asphalt cement concrete mixes, safety and signage, road reconstruction, cost efficiency, and many other areas.

More soils engineering and research on loads

Merlin Spangler became a first-rate soils engineer and performed useful investigations in this and related disciplines. Spangler collaborated with colleagues at Iowa State College or worked independently. The paper he co-authored for the 1940 Highway Research Board meeting examined wheel load stress on flexible pavements. Flexible pavements, particularly asphalts, are road surfaces that have "little or no inherent resistance to deformation under applied load." Spangler's laboratory experiments found that the pressure exerted by vehicles occurs at a relatively small contact area of the tire. He determined that the then-current formula correctly calculated the contact area between vehicles and flexible pavements and urged the field investigation of these matters.[85]

In 1941 Spangler theorized that moisture content of the subgrade related directly to the breakup of flexible pavements. He did not see it as the only factor but as the primary consideration. This moisture flow in subgrades was a topic of another paper he co-delivered that year, using Iowa State Highway Commission test data for support. He also began to theorize formulas to determine safe load levels on flexible pavements.

Spangler and M. B. Russell had previously published a paper on moisture flow in soil in an Iowa State College Agricultural Experiment Station journal. In 1941 Russell and Spangler proposed that conventional theory, an engineering theory called Darcy's law did not apply under certain situations. According to Darcy's law, moisture moves from high concentrations to low concentrations. However, Spangler and Russell found that if a textural gradient existed, such as a fine sand, the moisture would flow from low to high concentrations. With this research Spangler and Russell developed a new theory on moisture flow with applications for subgrade drainage.[86]

Spangler's 1942 study examined the design of flexible pavements. These roads consisted of four principal elements: a bituminous wearing surface, a stabilized base course, a subbase, and the subgrade or soil stratum. Spangler examined the deterioration of these pavements with continued focus on the subgrade. Spangler believed that the failure of these pavements resulted from extremely heavy loads that caused a type of breakup called "alligator cracking." Spangler developed a formula for pavement width that allowed proper deflection of the load and eliminated pavement failure.

Spangler's repeated discussions about flexible pavement failure raised the ire of asphalt paving interests. Two commentators on Spangler's paper rebuked him for grouping all flexible pavements into one class. Prevost Hubbard, representing the Asphalt Institute, and H. G. Nevitt, employed by the Socony-Vacuum Oil Company, disagreed with Spangler's generalizations. Hubbard and Nevitt asserted that bituminous surfaces like asphaltic concrete held up to heavy loads. In his reply, Spangler acknowledged these points but asserted that he classed asphaltic concrete as a semi-rigid pavement and not subject to the discussion.

His testing method considered only the truly flexible pavements. Spangler asserted that his formula proved reliable on flexible pavements.[87]

Spangler also continued to study the theory of loads on pipe. Marston and Spangler had developed what was known as the "Marston theory" or "Iowa theory" of loads on conduits. Spangler found that the theory could be adapted to flexible pipe such as corrugated pipe. Just as he had objections from asphalt contractors with his 1942 paper, Spangler now received some criticism from manufacturers and producers of flexible pipe products. G. E. Shafer of Armco Drainage and Metal Products, a corrugated pipe firm, commented that the Iowa tests had not gone far enough to determine the actual strength of flexible pipe. Shafer believed that additional tests should be conducted to determine the load level at which failure occurred.[88]

Spangler worked with O. H. Patel, an Iowa State College civil engineering student, on a study of southwestern Iowa's gumbotil soil, a plastic or sticky and heavy soil that challenged road builders in the state. The two presented this work to the Highway Research Board, which recommended methods for modifying the soil. Spangler and Patel sought to reduce the plastic nature of the soil by adding lime (CaO) or portland cement concrete. In their work, Spangler and Patel found that lime worked well as an admixture. The men advocated additional investigation, especially controlled field trials. Later examinations would determine that dolomitic limestone—limestone with magnesium—stabilized the soil better than limestone $(CaCO_3)$.[89]

Spangler assisted Harry King, another Iowa State College student from Anchorage, Alaska, with thesis research that they presented at the annual meeting of the Highway Research Board in 1949. Earlier researchers had postulated and demonstrated that electrical current passed between piling would increase the pile's bearing capacity. Spangler and King continued this research by studying the effect of electrical charges on clay soils. This method involved burying electrodes in the ground and passing a current between poles. Based on this research, Spangler and King made recommendations for the best practices to increase the bearing capacity of soils. The examination also showed that negative electrode piles outperformed positive electrodes for increasing the bearing capacity of the piling. The engineers also elaborated on an earlier study of soil bearing capacity and demonstrated the advantage of using negative electrodes.[90]

Merlin Spangler was not the only Iowa State College researcher interested in soils engineering. Richard Frevert, an agricultural engineer, and Don Kirkham, a physics professor, established a field method to test the permeability of soil. In Frevert and Kirkham's field tests in 1957, tubes of soil were pulled from the earth and a standard formula applied for measuring permeability. The results demonstrated that data from the field tests closely correlated to laboratory tests. This practice provided a practical alternative to the use of laboratories.[91] This information allowed engineers to test the suitability of soil at the site rather than in a lab, which is more economical if tests show the soil is so wet that it will not support a road.

Iowa State College Professor of Civil Engineering Donald Davidson examined methods to better stabilize soils and mixtures in road building. Davidson presented findings from two projects at the 1949 meeting of the Highway Research Board. In one study, he examined the soil stabilizing effects of six commercially produced products, such as Armac T

***Donald T. Davidson,
c. 1950***
*Iowa State College Professor
Davidson also investigated
soil engineering, including
methods of stabilizing soil.*

made by the Armour Chemical Company, and determined that some of these materials showed promise as stabilizing agents. Davidson advocated further investigation.

Davidson collaborated with John Glab of the Army Corps of Engineers on another paper at the 1949 Highway Research Board meeting, which similarly examined the relationship of organic compounds as stabilizing agents for soil-aggregate mixtures often used as subgrade materials. Davidson and Glab's results were similar to those in Davidson's paper on soil stabilization. The chemical agents showed promise in their ability to stabilize loess soil, but more investigation was necessary.[92]

Tracking costs

In 1942 Anson Marston chaired a Highway Research Board committee for the last time. He served as chair of the Committee on Uniform Accounting, and Robley Winfrey delivered a paper as part of the committee. Winfrey's discussion of accounting practices in Kansas that incorporated investment, mileage control, and road life in determining highway costs served as a model for highway accounting policy. Winfrey believed that all these items must be considered to determine the real cost of highways.

Winfrey advocated more accurate accounting practices in Iowa for the costs of highways, including tracking upkeep costs and salvage values, to provide engineers with better information on

which to base their decisions about highway construction. He cited the continued maintenance required on macadam roads as an example of how the useful life of impermanent surfaces could be extended with relatively little additional cost. The Kansas State Highway Commission had established standardized forms to accurately track roadway expenditures: investment per mile, investment per vehicle-mile, road cost per mile, road cost per vehicle-mile, revenue per mile, revenue per mile per dollar investment, and net earnings per mile on an annual basis. Such standardized record keeping, Winfrey believed, established credibility for the highway commission, helped engineers justify further development of highways, and helped the public understand the need for increased taxes to improve highways.[93]

War-time research

During World War II Ralph Moyer focused on issues relevant to war-time shortages. His 1942 study explored motor vehicle operating costs and the roughness and slipperiness of road surfaces. Moyer had begun a series of road tests in Iowa in 1938 and 1939 and wanted to investigate these issues on a larger magnitude. He conducted tests in 1941 and 1942 in Kansas, Missouri, and Wyoming on five types of bituminous surfaces and two types of concrete pavements. The tests monitored gasoline consumption, oil usage, tire wear, vehicle repair and maintenance, road roughness, and road slipperiness on a total of 450,000 miles of highway. The test team drove three Plymouth automobiles and included a support crew with a tow truck and two trailers.[94]

The results of the Kansas tests provide a brief synopsis of this extensive research. Two Plymouths were driven in these tests, one on bituminous surfaces only and the other on concrete road only. The vehicles were tested on the same day and time to minimize differences in the effect of weather. A driver and observer, who alternated positions every 50 miles, rode in each car. The teams also rotated between vehicles on a daily basis. The teams tested the vehicles at four different speeds: 35, 45, 55, and 65 miles per hour. To further ensure uniformity, a mechanic serviced each vehicle in a similar manner. The study addressed tire wear, fuel economy, and slipperiness in light of war-time shortages.

Moyer determined that tire wear was directly related to speed and the presence of sharp aggregates. This research supported the national speed limit of 35 miles per hour established during the war. Fuel and rubber were saved when speeds were reduced, allowing those resources to be used toward the war effort. Moyer suggested that to increase safety, a seal coat should be placed on bituminous surfaces. Although adding to the cost of the road, the seal coat increased traction.

Moyer was interested in the implications of his research for highway transportation in the post-war era. He addressed highway design and construction that would "contribute to lower operating costs, greater comfort, and greater safety." Regarding tire economy, he noted that road surfaces need not have hard, sharp aggregates included in the mix as a preventative to skidding, and he recommended that these aggregates be eliminated.

Moyer also believed that riding comfort, a primary concern of road users, should be improved through construction and maintenance methods, such as laying mosaic gravel texture bituminous roads, which are less rough than many bituminous surfaces. Although Moyer acknowledged the imminent construction of

four-lane divided highways, he asserted that two-lane roads would continue their predominance. Moyer's tests provided direction for highway engineers to design and build roadways that best served the public.

Portland cement and asphalt mixes

Researchers at the Iowa State College Engineering Experiment Station had been studying concrete mixes since the 1910s, and Iowa had become known for its portland cement concrete roads. This work and success, however, did not deter Iowa State College engineers from exploring the use of asphaltic concrete pavements. In 1949 Ting Ye Chu, an Engineering Experiment Station research associate, and Merlin Spangler presented a paper proposing a methodology to test asphaltic concrete mixtures. The Chu and Spangler test allowed plant supervisors to determine the general quality of pavement mixes at the plant rather than in the lab.

Chu and Spangler wanted a quick test that did not take the time that lab work required. Their test used a cylindrical mold that produced a specimen four inches in diameter and four inches high, the mold being slightly larger. The men compared the reliability of results of the field test to lab tests and the conclusion was favorable. Still in a preliminary stage, the men hoped to refine the test and make it applicable to other asphalt mixes. Whether these subsequent tests occurred is unknown.[95]

Expansion joints and aggregates

The Highway Research Board meeting in January 1946 allowed two Iowa State College faculty, Spangler and Moyer, a chance to discuss their expertise. Ralph Moyer continued his focus on pavement, with particular attention to the use of expansion joints. Expansion joints allowed the pavement slab to expand and contract with changing temperatures.

When portland cement concrete achieved common use as a road material, expansion joints were not always incorporated into the construction. The lack of allowance for pavement expansion caused many Iowa pavements to "blow up," especially in the summer months.

In 1934 Iowa State College had built a "large" stretch of concrete pavement. Moyer selected almost 500 feet of pavement in the area near the college highway materials testing laboratory and tested expansion joint material every 60 feet.[96] The joints consisted of nine unique materials: two of a standard asphalt mix and seven with compounds of cork, asphalt, and rubber. Iowa offered the necessary climate extremes to truly test the effectiveness of materials in expansion joints.

Iowa State College road test, c. 1938
Professor Ralph Moyer studied vehicle operating costs and road surfaces in Iowa. His tests monitored gasoline consumption, oil usage, and road slipperiness, among other things.

Although Moyer's report stated that the first year of testing had few profound differences in temperature, in 1936 the January low reached –36 degrees Fahrenheit and on July 15, 1936, the summer high reached 106 degrees Fahrenheit. Researchers measured the expansion in the winter and summer and found that the pavement absorbed moisture in the cooler months. While moisture in the pavement contributed to blowups, chemical and physical changes in the concrete, called "growth," created the primary problem.

The factors of time and high temperature created the situations in which blowups occurred. Moyer's personal observation indicated that most blowups happened with pavement aged 12 to 15 years, about mid afternoon on a "bright hot day in early summer." After a period of time, the materials in the joints lost their ability to buffer the expansion, contraction, and growth of the pavement that contributed to the blowup.

Moyer's study relied on data gathered by the Iowa State Highway Commission regarding coarse aggregates. In 1944 and early 1945 the Iowa State Highway Commission's Department of Materials and Tests surveyed selected sections of pavement composed of aggregates from different locations, with and without expansion joints. The survey noted that roads in Butler and Hardin counties made with limestone from the Alden quarry provided the most durable service. Ironically, pavements in Story County, home of the Iowa State Highway Commission, which included limestone from the LeGrand quarry, performed the poorest.

Moyer reflected on the use of expansion joints and noted the trade-offs in the practice. The Iowa State Highway Commission had reported, and Moyer observed, disintegration of pavement at the expansion joints and blowups that sometimes occurred even with expansion joints. Moyer concluded his report by commenting on the highway commission's investigations into pavement growth, or volume change, as he now called the phenomenon. The Iowa State Highway Commission hoped to evaluate aggregates and eliminate the need for expansion joints.

Improving winter road conditions

Ralph Moyer also chaired the Committee on Winter Driving Hazards for the National Safety Council while maintaining his position as research professor of highway engineering at Iowa State College. The Committee on Winter Driving Conditions had been organized in 1939 "to make a thorough investigation of the dangers peculiar to winter driving." The committee's primary objective was to develop data concerning traction on snowy and icy road surfaces and determine remedies for winter driving problems. The committee had reason to investigate winter driving; the death rate from accidents was 24 to 53 percent higher in winter than in the summer months.[97]

Moyer presented his last paper as an Iowa State College faculty member to the Highway Research Board in December 1947. The paper considered the effect of winter-weather driving conditions. Although Moyer did little of the actual research on his project, the study determined that the main dangers were inadequate traction and poor visibility. Snow and ice diminished traction, a situation that, when combined with poor driving technique, contributed to vehicle skidding. Visibility problems occurred in the winter due to three factors: increased hours of darkness, storms, and ice or frost on windows and windshields.

To approximate winter ice conditions, vehicles were tested on frozen lakes in Minnesota. Repeated traction tests were made to determine the effectiveness of bare tires versus tires with chains. The team also tested the efficacy of sand on iced roadways and the braking and accelerating ability of synthetic and natural rubber tires. The researchers scientifically determined that pumping the brakes of a vehicle decreased the necessary braking distance in a skid on ice, but the study offered few dramatic conclusions.

Iowa limestone quarry, c. 1950
Iowa State Highway Commission engineers investigated and recorded the quality of Iowa limestone for aggregates. Specific quarries and ledges were identified for the performance and durability of their aggregate, and the best aggregates were used in Iowa highways.

Signage

Other researchers from Iowa who were not highway engineers also contributed papers to the Highway Research Board. A. R. Lauer, professor of psychology and highway safety at Iowa State College, contributed papers on efficiency of signage. Lauer had conducted research on

No passing sign, 1958
The Iowa State Highway Commission promoted highway safety through increased signage. The commission first experimented with "No Passing Zone" pennants on Highway 30. The pennant is now a uniform traffic sign.

49

standardized license plates in 1930 and 1931. In 1932 he advocated standardization in highway signage to the Highway Research Board.

Lauer commenced studying signage at Iowa State College in 1940, and his work came to fruition at the Highway Research Board in 1947 with a report on his study of stop signs. Lauer advocated a change from the octagonal shape to a square or diamond because tests indicated that the octagon shape was not easily distinguishable. He also began preliminary tests of sign color. His models from Iowa and Ohio used black letters on a white background, but he concluded that reflectorized red and white would prove most distinctive. Lauer's advocacy for a change of shape was not adopted, but the color modification became a part of the standard stop sign.

Ralph Moyer, c. 1945
In the late 1930s and early 1940s Iowa State College researchers studied pavements and tire wear in Iowa and across the nation. Professor Moyer based his recommendations for dealing with shortages during World War II on this research.

A 1940 paper by Ralph Moyer focused on determining safe speeds and signage on highway curves. The project received support from the Highway Research Board, the National Safety Council, and the Engineering Experiment Station at Iowa State College. In the summer of 1940 Moyer and D. S. Berry, a researcher for the National Safety Council, began a systematic study of current practices to determine curve speed limits and signage. The research reported the findings of a questionnaire prepared by Moyer and Berry for each state engineer in charge of traffic control. The survey determined which states labeled speed limits for curves on state routes and how the limits had been established.[98]

The second half of Moyer and Berry's study sought to establish a standard method for determining a safe speed on curves. Although Missouri had conducted a preliminary study of safe speeds on highway curves in 1937, Moyer foresaw the value of a national standard, including a formula for setting maximum safe speeds on curves and a uniform marking system for curves. Signage would eliminate the surprise element for motorists, and standards established by engineers would be safer than judgments of individual drivers. Moyer emphasized the need for reflective signs because curves became more dangerous at night.

In an era when some states still had no authority to establish speed limits, Moyer and Berry's proposal faced bureaucratic inertia. Some engineers questioned why it was necessary to use signs when no significant problem existed. Others resisted what were perceived as attempts to limit individual freedom on the road. Some motorists believed speed signage on curves was obtrusive and restrictive. To combat such opposition, Moyer cited human safety, referring to an Indiana test in 1939 and 1940 where fatalities decreased by 10 lives, personal injuries by 12, and total accidents by 36 in the previous 12-month period after signs were posted on curves. Moyer asserted that public service and public safety were best served with this signage.[99]

Salvaging old pavements

After World War II the Iowa State Highway Commission concentrated on widening pavements. Vernon Gould, an engineer in the commission's Construction Department, reported on the process of salvaging old pavements.

The concept of salvaging referred to the practice of resurfacing over existing pavement as roads were widened. Beginning in 1948 Iowa experimented with this technique, using both portland cement concrete and asphaltic concrete. The Iowa State Highway Commission used primarily asphaltic concrete for the new surface but found both types of concrete to be economical and good engineering practice. Gould concluded his report with the commentary that when the Iowa State Highway Commission had begun resurfacing highways it consulted with other states like Ohio that had already used the practice for a while. He also noted that the Iowa State Highway Commission had conferred with equipment manufacturers to adapt existing equipment to meet resurfacing needs. Once again, the Iowa State Highway Commission promoted cooperation and investigation among states and private enterprise.[100]

Improving secondary roads

Although reported at a national level, L. M. "Slim" Clauson's paper on secondary road surfacing proved most valuable to Iowans. Clauson's report developed out of the Iowa State Highway Commission's Joint Board to Study Secondary Road Surfacing Problems. As the state and counties surfaced more secondary roads, materials became increasingly scarce, and the County Engineers Association appealed to the Iowa State Highway Commission for assistance identifying materials and their location. This appeal led to the creation of the joint board.

In response to the scarcity of aggregate, Clauson wrote, the counties began to experiment and improvise with previously neglected materials. Clauson explained that in all parts of the state county engineers experimented with various types of surfaces. In Muscatine in the southeast part of the state, a bituminous road responded well to truck traffic hauling gravel after the county stabilized the sand base. In the western Iowa county of Audubon, clayey sand was mixed with crushed limestone and compacted to create a satisfactory road.

Another practice that drew Clauson's attention was the process of reclaiming gravel and crushed stone from roads for re-use. This material provided a suitable base to surface low-volume secondary roads. Clauson and the joint board recognized the need to work with geologists and other professionals. Only through cooperation and additional investigation would new practices develop. The practices Clauson had reported demonstrated that both the counties and the state highway commission could be flexible. When materials became scarce, Iowa engineers and highway officials responded to the challenge.[101]

When Clauson wrote of the joint board he referred to an organization that turned out to be the predecessor to the Iowa Highway Research Board. The year 1949 saw two notable achievements for the Iowa State Highway Commission: the commission laid the first mile of slip-form paving in north-central Iowa, and in December it authorized the creation of the Iowa Highway Research Board.

Iowa's own Highway Research Board

The Iowa State Highway Commission organized the Iowa Highway Research Board in May 1950, and its first meeting focused on organizational matters. The 11 members elected William E. Jones, assistant to the chief engineer, as the board's chair and Bert Myers, head of the Department of Materials and Tests, as vice

Tile machine, Story County, Iowa, c. 1945
Subgrade drainage is critical to highway construction. If the subgrade becomes saturated with moisture, the highway will fail. Highway workers dug trenches and laid drain tile to move moisture away from the roadway.

chair. The commission had appointed Mark Morris to be the director of highway research as part of his duties as traffic engineer, and in 1953 he became the research director for the Iowa State Highway Commission.

The policy of directed highway research had come full circle. The state that had consistently provided the nation with highway engineering and highway research leadership had now established its own formal board to serve as a state

clearinghouse for road research. The first Iowa Highway Research Board consisted of 11 elected members, six county engineers, the dean of engineering at Iowa State College, the dean of engineering at the State University of Iowa (now the University of Iowa), and three Iowa State Highway Commission engineers.

The Iowa Highway Research Board continues to direct and recommend research today, but its early guidance did not have an immediate effect. Researchers would need several years to complete the initial projects proposed to the Iowa Highway Research Board. However, a larger pattern of state-supported research had been established, and Iowans received the benefits. By 1996 the Iowa Department of Transportation (successor to the Iowa State Highway Commission) had funded approximately 390 research projects submitted through the Iowa Highway Research Board.[103]

Mark Morris
Morris first worked for the Iowa State Highway Commission in 1915, and in 1949 he became the first research engineer for the commission.

Members of the Iowa Highway Research Board[102]

May 4, 1950 (term beginning January 1, 1950)

Name	Affiliation	Term
R. E. Robertson	County Engineer, Cerro Gordo County	One year
P. A. Michel	County Engineer, Montgomery County	One year
Walter H. Root	Iowa State Highway Commission, Maintenance Engineer	One year
C. A. Elliott	County Engineer, Greene County	Two years
J. R. Daugherty	County Engineer, Muscatine County	Two years
Bert Myers	Iowa State Highway Commission, Engineer for Materials and Tests	Two years
F. M. Dawson	Dean of Engineering, State University of Iowa	Three years
J. F. Downie-Smith	Dean of Engineering, Iowa State College	Three years
Edward Winkle	County Engineer, Osceola County	Three years
L. J. Schiltz	County Engineer, Dubuque County	Three years
William E. Jones	Iowa State Highway Commission Assistant to the Chief Engineer	Three years

Summary

During the 1940s Iowa engineers developed machinery like the slip-form paver to modernize and upgrade the state's highways. A farm-to-market road program was established, sometimes at the expense of primary roads, to provide better transportation and communication between rural communities. Researchers like Ralph Moyer studied safety and comfort issues to provide a better driving environment for Iowans, while Merlin Spangler continued the work on load theory begun by Anson Marston. The highway commission's dedication to research continued with the creation of Iowa's own Highway Research Board.

In the coming decade Iowans would participate with the rest of the nation in federal highway construction. Iowa engineers relied on their past experience and supported research to plan the federal system. During the next 25 years, Iowa's engineers would help create an important state and national road system.

Iowa Highway Research Board, c. 1959
The Iowa legislature authorized the creation of the Iowa Highway Research Board in December 1949. The board consisted of highway commission officials, county engineers, and university deans. It still meets to recommend funding for highway research.

Chapter 4

Building the Modern Highway: 1950–1974

After successfully devoting 35 years to hard-surfaced paving, the Iowa State Highway Commission faced three significant challenges from 1950 to 1974: the primary road system, established in 1919, required maintenance; the farm-to-market system demanded expansion; and, between 1954 and 1956, the federal government created the plan for a federally funded, four-lane interstate system.

Meanwhile, the long-time Iowa presence on the national Highway Research Board diminished when Roy Crum died in 1951 and Thomas H. MacDonald retired from the Bureau of Public Roads in 1953. For the first time in its 30-year history, an Iowa engineer did not serve on the executive committee of the Highway Research Board. The national Highway Research Board continued to promote highway research, and Iowa augmented that effort through the Iowa Highway Research Board.

During this period engineers from the Iowa highway commission and counties, along with the engineering deans of the Iowa state universities, met to recommend research funding priorities. Cooperation between Iowa State College and the Iowa State Highway Commission also continued after World War II. In addition, the University of Iowa, then the State University of Iowa, and the University of Northern Iowa, then the Iowa State Teachers College, increased their roles in highway research and engineering by working with the Iowa Highway Research Board.

The secondary road system

Iowa has a tradition of building quality highways, but the equitable disbursement of road funds among state and local agencies has sometimes been an issue. Bob Given, a former engineer at the Iowa State Highway Commission, says, "Originally, most of the road use tax funds were dedicated to construction of the state highway system, but eventually the Iowa State Highway Commission reached a point where the problem was one of saturation. We had state highways running to all the major communities and serving the principal needs of transportation interests, but we hadn't done a whole lot in the cities or in the counties. Dewey Goode, a legislator from Davis County, lobbied the highway commissioners for equitable relief."

In response, the commissioners revised programs, identified priorities, and reassigned some road use tax funds from the state highway system to the farm-to-market county programs and to the cities for their major urban arterial route needs. This equitable relief plan allowed a state-wide network of roads to be completed.[104]

Occasionally county engineers from Iowa reported the progress of secondary road work at the national Highway Research Board meeting. In 1950 William Behrens from Linn County in northeastern Iowa

Concrete finisher, c. 1925 (facing page)
Manually finishing the wet concrete was an essential part of road construction.

discussed secondary road administration. The Bergman Act of 1930 had taken jurisdiction over secondary roads away from township supervisors and given this jurisdiction to the county board of supervisors and county engineer. Behrens reported on the county road department and its financial responsibilities and about the sources of revenue for secondary road work.

According to Behrens's report, Iowa counties used property taxes, local assessments for road surfacing, and state road use taxes to construct and maintain county secondary highways. Linn County operated a central maintenance shop to house machinery and repair equipment. Behrens stated that grading earthen roads "is an expensive item," but the county reduced its costs by surfacing more roads. Much of the maintenance involved surfacing and resurfacing roads with stone, and the county applied calcium chloride to minimize the amount of grading necessary.

Behrens also commented on the value of aerial maps to county engineers. The Agricultural Adjustment Administration generated these maps and, although Behrens did not explicitly state their uses, he wrote that "it would now seem impossible to function properly without them." Presumably the county officials could plan and implement their highway program more effectively with these maps. Behrens believed Linn County's use of aerial photos to plan its road program reflected the quality of secondary road administration in Iowa.[105]

The slip-form paver also provided rural Iowans an opportunity to truly get out of the mud. In 1962 Pocahontas County in northwest Iowa became the first county in the nation to have more than 100 miles of slip-formed portland cement concrete road surfaces. Other Iowa counties used the paver on lower volume roads, and the

Iowa State Highway Commission used slip-form paving on the interstate system. The commission experimented with various thicknesses of pavement, and on low-volume secondary roads six inches became the standard. A 1969 article in *Better Roads* reported that the slip-formed pavements of Greene County laid in 1951 had "pleasantly surprised" the county engineer. Slip-form paving benefited all Iowa constituencies.[106]

The interstate era begins

With the Federal Highway Act of 1956, America entered a new age of highway building. The act appropriated $24 billion over 13 years for the construction of a limited-access, interstate highway system. On September 21, 1958, the first section of federal interstate in Iowa, a stretch of Interstate 35/80 southwest of Des Moines, opened for traffic. The interstate system would eventually extend 781.51 miles across Iowa, but the first section to be built in its entirety was the short link of Interstate 480 between Council Bluffs and Omaha. In 1985 the interstate system was completed when Interstate 380 from Iowa City to Waterloo was finished. As traffic demands increase, additional research and planning continue to be needed to support interstate maintenance and plan new routes.[107]

Problems with federal standards

The federal government established standards for building the interstate system that occasionally concerned Iowa State Highway Commission engineers. One former commission engineer suggests that the states had better judgment. Don McLean, a design engineer for the Iowa State Highway Commission when the interstate highways were being built, believes the federal government placed undue emphasis on highway design formulas created by Wally Little, a federal government engineer, who had developed his formulas after tests on an Ottawa, Illinois, test road.

Based on these tests, the federal government dictated eight-inch pavement thickness formulas for the interstate system. The experiences of Iowa state engineers, who knew the local situation better, did not support the wholesale use of eight-inch pavement. They believed that on a few sections of interstate a higher standard would be more economical, but the federal government supported only the construction of eight-inch pavement. The federal guidelines allowed little flexibility in the plan for the interstate system; fortunately, instances of disagreement between state and federal officials regarding design standards were rare.

To build 10-inch thick interstate highways at all, Iowa State Highway Commission engineers had to justify the higher standards to highway commissioners and legislators, who were not eager to spend additional funds for 10-inch pavements when the federal government said eight inches would do. The design technician and the design engineer would sometimes have to settle for the federally approved eight inches in spite of their better judgment. A former design engineer reflected that federal specifications stifled innovation because engineers were forced to settle for the minimum standards.[108]

A specific instance of the arbitrary nature of federal standards for highway construction is demonstrated in bridge design. Several states tried to build wider bridges than federal standards dictated because many engineers believed that a bridge only two feet wider than the pavement was too narrow. From a safety standpoint, engineers argued that bridges should be a full shoulder width wider on each side than the travel lanes, that is, about 22 feet total. But if the state highway commission built the extra width, state taxpayers paid for the addition without federal assistance. In some instances, the federal

programs adopted new standards, making all bridges under construction below standard before they were even finished.

Iowa engineers experienced similar difficulties with guidelines for bridge and overpass clearance. Federal standards dictated a 14-foot clearance so that trucks could pass under bridges or overpasses. Design engineers adjusted the grade to provide for the 14-foot standard, but engineers often struggled to meet those specifications. As designers worked to engineer the grade down to achieve the 14 feet of clearance, the federal standards were arbitrarily changed to require a 16-foot clearance. Such situations occurred only rarely, but when they did Iowa's engineers were understandably frustrated.[109]

Peat in northern Iowa

Federal guidelines were not the only source of problems for the Iowa State Highway Commission as it constructed interstate highways. For example, the soil in the area of Interstate 35 in north-central Iowa contained a considerable amount of peat, which was unsuitable as a road bed because it did not provide adequate support. Gus Anderson comments that in some areas the peat reached a depth of 35 feet.

To counter the problem, soils engineers developed a plan to overload the road bed by putting "borrow materials" (that is, suitable soil removed from an adjacent area) on the road bed. The borrow materials did indeed create a sound bed, but the tremendous pressure of the fill on the peat caused the peat to squirt out the side of the roadway, affecting nearby utility poles. Anderson explains, "The telephone poles would tilt in all directions as we squished that muck out." The next attempt to stabilize the road bed involved placing about three feet of extra fill above the final gradient, then remov-

Construction of a borrow pit, 1967
Borrow material—soil—from pits is used to raise the grade of a road. Today ponds are located near many highway bridges and overpasses where borrow material was removed.

ing it before paving the road. This technique stabilized the soil, and when the paving was placed it did not push the subgrade down farther.[110]

The Iowa State Highway Commission developed another method to address the problems created by the peat soils in north-central Iowa. The contractors removed the highly organic topsoils and left the stable glacial soil in place.

Don McLean, former design engineer and assistant chief engineer at the commission, explains the removal of the boggy soils: "On later projects, where we'd run into peat-like soil, we went in with equipment and bailed it out. We placed it off to the side of the road, and there was quite a bit of that." Creative solutions in soil engineering allowed the work on Interstate 35 to proceed.[111]

Steel shortage

Another serious problem associated with building Iowa's interstate highways developed late in the 1950s when steel became scarce and delivery times were delayed, a costly development that the Iowa State Highway Commission simply could not afford. Fortunately, the Iowa Highway Research Board had been supporting experimentation with new materials and techniques for constructing interstate highways. In response to the steel shortage, the highway commission looked at a possible substitute material, aluminum.

The commission worked with former University of Iowa engineering professor Ned Ashton, aluminum producers, and private contractors to construct a welded aluminum highway overpass on 86th Street in Urbandale, also called Clive Road, over Interstate 35/80. Jensen Construction Company and United Contractors of Des Moines submitted the low bid for the project. ALCOA, Kaiser Aluminum, and Reynolds Metals participated in the work by providing $10,000 each and supplying information on aluminum alloys. Spanning 222 feet, the overpass had a deck 36 feet long and 30 feet wide with two-foot sidewalks on each side. The structure was opened on September 24, 1958, and performed well until it had to be replaced in the early 1990s as part of a new interchange project.[112]

The Clive Road bridge project demonstrated that aluminum provided some advantages over steel. Specifically, aluminum did not corrode and had a shorter delivery time. One article describing the project speculated that aluminum might be able to compete with steel for use in highway construction. But the project was the only one of its kind in the nation. The aluminum cost $47,000 more than for an equivalent bridge, and in the early 1960s steel again became easily obtainable. Additional plans to build welded aluminum bridges were never considered.[113]

Safety issues

Yet another serious problem with the interstate highway system involved safety. Limited access to interstate highways made it difficult for public officials to respond to emergencies. The Iowa State Highway Commission responded by putting new technologies to work to ensure motorist safety. In 1973, for example, the Iowa State Highway Commission and the Iowa State Highway Patrol cooperated to establish the Highway Emergency Long-Distance Phone—HELP—system. Motorists could use a toll-free phone number to report emergencies along Iowa's interstate and primary highways to the Highway Patrol in Des Moines, and a dispatcher notified the appropriate local authority to assist the caller. A report about the system's first six months of service showed that many callers desired weather information, a peripheral service of the HELP line. Callers also reported accidents or unsafe road conditions to the dispatcher.

One caller who telephoned to request assistance when his car ran out of gas got more help than he bargained for. The state trooper who picked up the stranded motorist and his companion conducted a license check on the vehicle, determined the car had been stolen, and took the two men to the Grinnell jail. In more ways

than one, the HELP system demonstrated the Iowa State Highway Commission's continued emphasis on motorist safety.[114]

Toward better roads

While addressing the needs of the interstate and secondary road systems, Iowa's highway professionals developed better planning and design methods, continued their research of soils and aggregates, and improved maintenance practices.

Studying scour

The Iowa Institute of Hydraulic Research at the State University of Iowa in Iowa City increased its research presence in highway engineering through the Iowa Highway Research Board. The institute had become one of the nation's earliest hydraulic research facilities after the university completed construction of its laboratory in 1919.

During the 1950s the institute received support to study scour, that is, the strength and depth of the force of water around bridges and abutments. The first study investigated four features of water and bridge piers: the geometry of piers, stream-flow characteristics, sediment

Aluminum bridge, 1958
The first and only aluminum highway overpass was opened over Interstate 35/80 in the Des Moines area. When steel became subject to delivery problems in the late 1950s, the Iowa State Highway Commission experimented with the use of aluminum as an alternative building material. After steel again became more available in the 1960s, the highway commission did not use aluminum in bridges again.

Steel inspection, 1975
Materials inspection has always been of critical importance to the Iowa State Highway Commission. Steel for highway construction can be inspected at the plant or at laboratories.

characteristics, and the geometry of the channel. In part, institute researchers Emmet Laursen and Arthur Toch determined that a streamlined pier of the smallest structurally reliable size—the size that used the least material and still guaranteed structural stability—was the preferred design. They also determined that in flood conditions, or other times of uneven flow, the scour depth increased.[115]

The institute followed Laursen and Toch's report with a paper by Philip Hubbard at the January 1955 meeting of the national Highway Research Board. Previous reports on bridge-pier scour relied on laboratory research rather than field tests, but Hubbard's paper reported on field measurements conducted by the institute. The paper discussed the institute's goals to establish a model-prototype for idealized situations, to study the model in more complex situations, and to determine the model's reliability for applications.

For more than a year Hubbard field-tested on the Skunk River a gauge apparatus developed at the institute. The results served as a basis for studies on additional streams and rivers. Hubbard found that highway commission laboratory employees could monitor the device with minimal instruction. He believed

that any engineer or engineering assistant could be trained to use it. The next step would be wider field tests of the gauge.[116]

A paper presented by Emmet Laursen in January 1955 at the annual meeting of the national Highway Research Board reported results from a more complicated trial on the Mississippi River near Dubuque. Laursen concluded that the results of the Mississippi River test, which paralleled those of the Skunk River test, indicated that sediment in the stream played a significant role in scour. The effect of sediment was difficult to duplicate in the laboratory and, although the general pattern of scour was similar, the details differed between lab tests and field tests because of the difference in scale and the presence of suspended sediments in the field tests. Laursen's report indicated that a general relationship between field and laboratory tests existed, but additional trials were necessary.[117]

Foamed asphalt

The year 1949 proved to be a landmark year for both the highway commission and Iowa State College. During that year the legislature established the Iowa Highway Research Board, and the college hired Civil Engineering Professor Ladis Csanyi. A transplant from New York state, Professor Csanyi acted as the chief researcher for the new Bituminous Research Laboratory at the college, directing work that proved useful for building roads on a local and international level.

For example, Csanyi's work in foamed asphalt, a process of injecting steam into asphaltic cement, received international attention. Foamed asphalt provided an inexpensive method for paving less traveled roads and parking lots. Beginning with a 1953 trip to the Federal Republic of Germany, in little more than a decade Csanyi had visited all the continents except Antarctica promoting this surface. One of his more publicized trips occurred in 1962 when Csanyi ventured to Australia to advocate the use of foamed asphalt as a suitable surface for road construction in the Outback.[118]

Csanyi's work on foamed asphalt grew out of the need to improve the secondary road system. Maintenance costs on crushed stone roads proved expensive, and the clamor for hard surfaces continued. In response to the desire for higher quality and affordable secondary roads, the Bituminous Research Laboratory at Iowa State College developed a method to stabilize road bases with foamed asphalt. Together with Iowa State Highway Commission researchers, laboratory personnel selected a trial road, 24th Street in Ames, and in 1957 prepared it for the application of the foamed asphalt base.

The road had received little maintenance since 1951. Prior to treatment with the foamed asphalt, a grader smoothed the road to a uniform three-quarter-inch gravel surface, then a scarifier performed another pass to mix up the existing road materials, and finally the prepared surface was moistened. The foamed asphalt was applied by a Seaman-Andwall Pulvi-Mixer, a commercial tractor-like machine with a rear-mounted sprayer bar with 12 nozzles to apply the asphalt. The mixer had a steam boiler attached to its front and towed an asphalt supply kettle with an outrigger alongside the machine. The mixer applied the foamed asphalt in four passes to obtain the desired amount of material. A roller compacted the subgrade, and a seal coat was applied to the pavement. After the pavement set, cores removed from the road were subjected to freeze-thaw tests, which demonstrated increased durability. The tests determined that a durable foamed asphalt cement base could be created economically, but researchers recommended improving the equipment to apply the foamed asphalt in one pass.[119]

Planning and controversy

In the 1950s the Iowa State Highway Commission had not yet formalized long- and short-term planning procedures.[120] Although Chief Engineer Fred White had proposed informal multi-year road plans as early as the 1920s, one-year construction plans were still the practice, with the continued emphasis on hard-surfacing primary and farm-to-market roads. Former Iowa State Highway Commission engineer Gerhard (Gus) Anderson, remembers enlisting the aid of a farmer in 1951 to rescue a commission survey vehicle stuck in the quagmire that passed for a road. The commission still had a long way to go to fulfill its long-time promise to "get Iowa out of the mud."

Ladis Csanyi, c. 1960
Csanyi directed research at Iowa State College's Bituminous Research Laboratory. He developed a product called foamed asphalt, which is still used worldwide.

At the same time, the federal government was forging ahead with plans for an interstate system of four-lane highways. With the construction of the interstate system increasing the demands on Iowa State Highway Commission resources, a long-range plan for building roads became imperative. In 1959—a year after the first interstate section of road was built in Iowa—the Iowa legislature first mandated that the highway commission conduct long-range planning. The commission's 1960 annual

Pavement core drill, 1966
As highway construction expanded, testing the durability of pavements increased. In 1966 the Iowa State Highway Commission acquired a portable pavement drill, which facilitated testing of portland cement concrete.

**Project planning,
c. 1969**
*The Office of Planning
coordinated the
Iowa State Highway
Commission's highway
plans beginning in 1959.
Planning engineers initiate
Iowa highway policy.*

report detailed the creation of the Division of Planning to comply with this mandate. The commission also provided for an engineer of planning who reported to the chief engineer and directed the Traffic and Highway Planning Department, the Secondary Road Department, the Urban Department, and the Highway Research Department.

The new planning process used a priority system already in operation but established a five-year plan to be updated annually. As part of the work of establishing a five-year plan, the Iowa State Highway Commission held public hearings. These hearings received increased attention and attendance because of public interest in the interstate system and the state of Iowa's purchase of rights-of-way for that system.[121] The commission also prepared the *Iowa Highway Needs Study* to forecast future demands on Iowa roads. Completed in 1960, this study served as a model for future maintenance and construction of Iowa's interstate, primary, and secondary roads.[122]

Occasionally someone attending an Iowa State Highway Commission public hearing became unruly. Former Iowa State Highway Commission engineer and Iowa

State University professor emeritus Stanley Ring recalls one memorable meeting in southern Iowa. A disgruntled service station owner, whose access to his station was limited by a road improvement, read a statement of discontent into the hearing record. After the meeting, he confronted Ring with his grievances.

Ring recalls, "After the meeting, I was talking to the city engineer, and this guy came up and started talking to me, using a lot of swear words. The city engineer in a nice way told him, 'You shouldn't be talking to him that way. Not only is he a friend of mine but he is doing his job.' And the guy turned around and hit him and dropped the city engineer, and they got in a big fight until they were separated." Instances like this were rare, but occasionally tempers flared over highway policy.[123]

Improving pavement design

During the 1950s the design of pavements, particularly material components and thickness, was generally decided by committee based on traffic volume and type of pavement and relying primarily on past experience rather than systematically gathered data. Don Anderson, former soils engineer for the Iowa State Highway Commission, recalls, "It was like, so, how thick should we make this pavement?"

At times, the research on vehicle weights was basic raw data instead of truck volume and axle weight data. Anderson explains, "The planning rarely relied on tests of potential load or a determination of subgrade soils which supported the pavement. There was a lot of work to be done."

The Iowa State Highway Commission began a process to formally investigate and research pavements. Anderson comments, "I had an interest in research from day one because I felt that a lot of benefits could be made in reducing pavement thicknesses and using higher quality paving materials." As a soils engineer, Anderson reviewed research literature on pavement thickness to identify better methodologies to design highways. In the construction of modern highways, the Iowa State Highway Commission relied on research generated through the Highway Research Board in addition to the commission's own research.[124]

During the 1950s pavement thickness and materials received additional attention with the construction of interstate highways and state roads, and the importance of research to improve pavements became increasingly evident as Iowa State Highway Commission engineers sought to improve the quality and durability of federal and state highways.[125] Before then, Iowa State Highway Commission engineers had based pavement thickness on traffic volume, a method that had proved adequate. The design engineer and the deputy chief engineer decided the road thickness for portland cement concrete roads and asphalt pavements. Eventually the highway commission utilized the Portland Cement Association design methods and Highway Research Board specifications developed in the late 1940s for portland cement concrete roads.

Improving asphalt pavement design
For many years Iowa relied on portland cement concrete for primary highways and asphalt pavements for lower-traffic roads. In spite of asphalt's relatively lower cost, the highway commission was reluctant to use asphalt designs for more heavily traveled roads, perhaps because of poor performance of some asphalt pavement designs on the more heavily traveled roads. To be competitive, asphalt pavements needed to be developed that could tolerate any load.

Because flexible pavements rely on the subgrade to absorb the load, better analysis of the foundation soils proved valuable in improving the load capacity of asphalt pavements. As work on the interstate highways progressed and primary and secondary road construction continued, soil research remained important and created a larger body of knowledge for engineers to consult as they designed roads.[126]

The base material was also critical for durable asphalt roads. For a long time the Iowa State Highway Commission specified a basic rolled-stone base with up to three inches of asphalt as the wearing course. The rolled-stone bases were untreated and easily became saturated with water, particularly in the spring and fall because of Iowa's humid climate. The moisture caused the roads to swell and soften, leaving the pavement with little strength and causing it to fail.

The commission also experimented with a bituminous-treated base design for asphalt roads. The aggregate particles in bituminous-treated bases were coated with an asphalt mixture, which created a bond and greater strength in the base. This treatment resulted in better performance than the rolled-stone, untreated bases. The commission engineers initially failed to recognize that the gravels with bituminous treatment had more inherent strength under saturation than the untreated rolled-stone bases. After researching the rolled-stone and bituminous-treated methods, however, and evaluating the performance of each design in equal situations, the Iowa State Highway Commission soil engineers determined that bituminous-treated bases outperformed the untreated rolled-stone bases, and the commission converted to the use of bitumen as a binder for all types of aggregate. The commission required this asphalt treatment, both with gravels and crushed limestone bases.[127]

Pavement life program

To construct long-lasting, durable highways, Iowa State Highway Commission engineers developed a target period for pavement life for program design purposes. The commission decided on 20-year pavement life programs. Bob Given, a former Iowa State Highway Commission engineer, explains: "That is not to say that in twenty years there would be a failure. The twenty-year concept was a matter of durability and capacity, not strength. The planners believed they had information to say that for any Iowa State Highway Commission road, it could carry traffic for twenty years." The Iowa State Highway Commission postulated that portland cement concrete, because of its proven strength, would last an additional 10 years.

Establishing weight limits

Design engineers were designing roads so that the load would be distributed into the subgrade, a method that provided maximum road performance.[128] The concept of loads concerned the distribution of traffic weight and its application to the pavement. Highway building plans relied on estimates of traffic volume based on a certain percentage of trucks, population expansion, and industrial growth. The concept of road service considered both vehicle weights, which depended on weights allowed by law, and the distribution of weight, which depended on axle spacing and the number of tires in contact with pavement.

The Iowa legislature has traditionally relied on the state highway commission to recommend maximum limits for truck weight and to enforce the standards after they become law. As traffic volumes increased through the years, the commission changed its method of determining maximum truck weight standards. Through the 1930s the Iowa State Highway Commission had relied on gross weight as a means to establish maximum

weight limits. Then in 1941 the Iowa State Highway Commission began to limit weight by axle, implementing a weight limit for single axles at 17,000 pounds and, in 1945, setting weight limits for tandem axles.

To provide better information for road construction, the Iowa State Highway Commission needed more data on actual truck volumes and axle weights. Traffic engineers compiled the data, and design engineers applied the information to pavement performance. Using more accurate truck and traffic data allowed designers to maximize the durability of highways. Early traffic volume and weight research remained important for building the highway system when successive federal-aid highway acts relaxed truck weight standards beginning in 1974.[129]

More soils engineering

Donald T. Davidson continued to research Iowa soils, concentrating on the loess soils of the Wisconsin drift in western Iowa. Davidson reported on two projects in 1952. Davidson and Ting Ye Chu, a research assistant at the Iowa State College Engineering Experiment Station, examined the loess soil in coordination with the Iowa State Highway Commission. The commission was seeking a method to disperse the soil, or break it into its various components, to examine the properties of loess soil for highway construction.

The Iowa Highway Research Board recommended this project for funding, and the research resulted in a paper at the national Highway Research Board meeting. The study drew the interest of the Iowa State Highway Commission because almost two-thirds of the state was covered by loess soil. The dispersion test developed for loess soil allowed engineers to analyze soil and determine its suitability for road building.[130]

As part of the larger examination of the loess soil of western Iowa, Davidson studied the amount of clay present in Iowa's loess soil. Along with John Sheeler, another research associate at the Engineering Experiment Station, Davidson hoped to establish a relationship between the clay fraction and moisture content of the loess soil. The researchers examined more than 100 samples of soil and confirmed the direct relationship between clay and the effect of moisture on the loess soil. As the presence of .002-mm clay particles increased, the soil became more slippery, among other problems. The overall analysis, however, supported the findings of an earlier British study that no relationship existed between the percent of moisture in loess and the actual clay ratio.[131]

Davidson served as an investigator on three projects discussed at the 1957 meeting of the national Highway Research Board. Two research assistants at the station worked with Davidson to analyze stabilization of loess soils in western Iowa. The first study investigated the treatment of subgrades with a chemical Arquad 2HT manufactured by Armour and Company of Chicago. In a preliminary investigation, Arquad 2HT had shown potential to stabilize clay and loess soils. The 1957 test demonstrated that the chemical waterproofed the clay and made the soils more physically resistant to moisture and temperature changes. The study did not report on the cost of the treatment.[132]

Davidson assisted with an investigation of the treatment of loess with a resin from the combination of the organic liquids aniline and furfural. The resin produced moisture-resistant properties in the soil, making it waterproof. His report acknowledged the prohibitive cost of the treatment and gave particular attention to the health hazards of aniline, which caused health problems ranging from nausea

State fair exhibit, c. 1955
Because of the federal-aid interstate highway program, the Iowa State Highway Commission further promoted public education. The Iowa State Fair served as one venue; public hearings were another.

and headache to coma and cancer. Furfural, though not harmless, caused less severe problems, such as inflammation of the mucous membranes, after considerable exposure. For these reasons the treatments never gained widespread acceptance.[133]

Iowa State University demonstration of prestressed concrete beam strength, c. 1955
The engineering laboratories continued to serve as research and educational facilities throughout the 1950s.

Davidson collaborated on an additional loess stabilization investigation with L. W. Lu of Lehigh University to study the effect of lime on loess soil. The researchers investigated the performance of quicklime (calcium oxide) and dolomite, which

Collecting loess samples, c. 1955
In the 1950s Iowa State College researchers gathered loess soil samples from cutaway banks to investigate the properties of Iowa soil. The researchers wanted to find a means to stabilize loess for its use as a base material.

differed from quicklime because of the presence of magnesium. The researchers found that soil treated with the dolomitic quicklime resisted moisture better than soil treated with traditional quicklime. The use of dolomitic quicklime doubled the stabilization factor. This soil stabilization process provided an additional option to engineers building roads in western Iowa.[134]

Merlin Spangler's work on soils came to fruition in 1951 with the publication of his book, *Soil Engineering*, a textbook aimed at upperclass undergraduates and engineers without formal training. Commented Spangler in the preface to the text, "[S]oil is an engineering material and needs to be studied and handled in an engineering manner," reflecting his attitude, developed during more than 20 years of research, about the importance of soil in highway engineering and other engineering applications. He included occasional references to pre-20th century developments in soils work and the impact of land forms and soil on society.[135]

Spangler's book was extensive, covering soil engineering in 25 chapters, and proved a useful reference. Spangler's inclusive treatment of frost heave provides an example of his comprehensive work. Although frost heave was not an area of specialization for Spangler, he advised engineers that the formation of "lenses," or layers of ice in the subgrade that broke up many expensive hard-surfaced roads every spring, caused the heave.

To solve the problem of frost heaves, Spangler suggested three techniques. Engineers could lower the level of the water table in the subgrade by using drain tile. They could use a coarse-grained subbase, such as crushed stone, which would reduce the capillary action that created the lenses. The third method was to use a clayey "cut-off blanket" beneath the pavement. Clay has great capillary potential, but water moves slowly through it, inhibiting the formation of ice lenses. Spangler's *Soil Engineering* became a standard in the engineering profession. After 40 years Spangler's work still serves as a resource for undergraduate engineering and Iowa's highway engineers.[136]

Aggregate research
Since Samuel Beyer had joined the Engineering Experiment Station staff at Iowa State College as a mining engineer in 1905, Iowa State College researchers had been investigating the location, and later the durability, of road materials. Ralph Moyer reported poor performance of aggregates at the 1945 meeting of the Highway Research Board. The Iowa State Highway Commission and Iowa State College had determined that pavement composed of limestone aggregate from the LeGrand quarry deteriorated faster than other pavements.

At the 1955 Highway Research Board meeting, Iowa State College geologists presented information on the relation of geological factors to the quality of aggregates. Iowa State Highway Commission engineers speculated that the distress to the pavement resulted from an inherent flaw in the stone, but they had not systematically analyzed the properties of limestone aggregate.[137]

LeGrand limestone came from east-central Iowa in Marshall County. Pavements made from this stone experienced failure due to a deterioration called D-cracking. D-cracking is characterized by a series of fine cracks in the pavement. In 1953 and 1954 Chalmer Roy, geology professor at Iowa State College, led a team of Iowa investigators in the analysis of Iowa limestones. The researchers examined stone from Fort Dodge and LeGrand quarries for the calcite-dolomite ratio and clay minerals. The researchers performed a thermal analysis of the stone and examined the stone with X-rays. The study examined sound and deteriorated concrete but found no distinguishing characteristics that caused D-cracking in pavement.[138]

The search for quality aggregates continued into the early 1960s. Don Anderson, a former Iowa State Highway Commission engineer, comments that finding aggregates was "a big problem because in highway construction, about half the total cost of a roadway is in the pavement. That is where most of the outlay is, and when you have premature deterioration of the pavement, whether it be an asphalt pavement because it had a rolled-stone base, or a concrete pavement because it had an aggregate that wasn't durable, that was a road built with public funds depreciating too rapidly."

Finding quality aggregates, therefore, was an important task. Anderson elaborates, "It was important to predetermine when a new source or new ledge of aggregate was going to stand up under the conditions in Iowa: using salt to increase traction on icy roads, and the freezing and thawing cycles. So each pavement type had its problem." The identification of suitable road building materials remained a priority for the Iowa State Highway Commission, and today's Iowa Department of Transportation retains records on aggregate performance to guarantee quality performance from material.[139]

The crushed limestone of Iowa's secondary road system also offered a persistent problem to engineers and citizens. The dust from the roads constructed from this material annoyed drivers. In northern Iowa in 1952 and 1953 the Butler County engineer experimented mixing sodium chloride and calcium chloride with clay as a binder on crushed limestone. Then, throughout the summer, work crews applied additional chemical to the road surface. The county engineer selected sodium chloride, due to its cheaper cost, to stabilize the base course. The engineer acknowledged the threat to water supplies from the chemical treatment of the material but proceeded with the project. He reported a reduced need to blade roads and the satisfactory performance of the roads. Because the experimentation was still in its infancy, the engineer commented that the true cost-benefit could not yet be determined.[140]

Resurfacing roads: a practical alternative to total reconstruction

When Iowa began resurfacing highways, the Iowa State Highway Commission was unsure of the usefulness of this program. To determine the effectiveness of resurfacing, the commission systematically studied the performance of previous work. In 1955 Laboratory Chief James Johnson reported on the progress made in resurfacing Iowa highways to the national Highway Research Board. He had studied 100 miles of road that had been resurfaced between 1931 and 1954. He specifically examined the performance of the pavements based on a five-year grouping. Resurfacing projects from 1931 to 1935 comprised the first study, with the other projects placed in five-year blocks up to the year 1950. The projects for 1951 to 1954 were listed individually.[141]

The commission researchers found a few problems with the resurfaced roads. In one instance in northwest Iowa, reinforc-

James Johnson, 1961
Iowa State Highway Commission Laboratory Chief Johnson was instrumental in developing the slip-form paver. He also reported commission investigations to the national Highway Research Board meetings.

ing mesh had been placed too near the surface of the new pavement, which created spalling—the breaking or chipping of pavement at joints, cracks, or edges. Pavement on a southeast Iowa resurfacing project also spalled where dowel bars had been placed too near the surface at an expansion joint.

Johnson reported that the rest of the resurfaced roads were in good to excellent shape, although he could not conclusively evaluate the performance of the newer surfaces on deteriorated original pavement. On a 21-mile section of Highway 30, where Iowa State Highway Commission maintenance crews had frequently patched sections of shattered pavement,

Road resurfacing, 1965
In addition to portland cement concrete, the Iowa State Highway Commission paved roads with asphalt. Workers used asphalt as an overlay on older highways.

the new surface remained free of the problem. However, the resurfaced section was less than 10 years old, not yet the age at which most pavements began to exhibit deterioration. He also referred to the performance of a resurfaced area on another part of Highway 30 near

Marshalltown constructed with inferior aggregate. The latest surface performed satisfactorily. Johnson surmised that resurfacing work with substandard aggregate outlasted a new slab of pavement of the same material, but the research was inconclusive on this aspect as well.[142]

Maintaining secondary roads

The maintenance of secondary roads received increased attention nationally. In 1960 the National Association of County Engineers conducted a study to formulate standards for secondary road maintenance and created a survey to determine county maintenance practices. The survey examined surface blading, weed-control, signage, bituminous pavements, centerline striping, and bridges.

Melvin Larsen, Iowa State Highway Commission secondary road engineer, reported to the 1961 meeting of the national Highway Research Board that 94 of Iowa's 99 counties responded to the survey. The responses indicated significant differences among counties in their emphasis on highway maintenance. Larsen hoped that county engineers could establish formal maintenance schedules based on the surveys. He believed that formal planning gave county engineers flexibility in emergency maintenance situations. If a snowstorm struck and an unusual amount of road work was necessary, for example, county engineers would better know how to manipulate their budgets to meet the emergency. Furthermore, uniform maintenance planning and practice by county engineers provided similar driving experiences for motorists across the state. Systematic, uniform maintenance procedures among Iowa's counties served taxpayers and road users.[143]

Partial-depth repair of bridges

Long-term maintenance of roads and bridges became increasingly difficult as the highway system aged. The annual winter cycle of icing on bridges had long proved dangerous to Iowa motorists, and midwestern highway departments mitigated the hazard by applying salt. Engineers at the Iowa State Highway Commission discovered that the salt brine penetrated the concrete and caused the deterioration of bridge decks, the concrete wearing surface of the bridge. Eventually the reinforcing steel bars, re-bar, of the bridge degraded as well, and repair proved costly.

Rather than replacing the entire bridge, the highway commission experimented with placing a new one-inch portland cement concrete wearing surface on the bridges—a partial depth repair rather than a complete resurfacing. The first test project was conducted in 1964 on a 240-foot bridge in northwest Iowa in Sac County. The old deck was scarified, or broken up, to a quarter-inch depth, and unsound concrete was removed with a jackhammer and a chipping hammer. Finally, workers removed the concrete around the reinforcing bar, and the re-bar was sandblasted to clean it. Work crews poured a low slump portland cement concrete (one with relatively low water content so it would set up faster) on the freshly prepared surface, then spread and smoothed it. After the concrete cured, the bridge was re-opened to traffic. This process is now universally accepted as an economical and sound method of bridge maintenance.[144]

Summary

In 1974 the state of Iowa reorganized the Iowa State Highway Commission as part of the new Iowa Department of Transportation. The engineers and staff of the Iowa State Highway Commission had served the state and the nation nobly and well for 70 years, providing Iowans with the hard-surfaced roads demanded throughout the first half of the century, and providing the country with valuable highway innovations and research.

Bridge deck removal, 1974
Iowa State Highway Commission engineers developed a process of bridge deck maintenance that came to be called the Iowa method. The Iowa method was a process of resurfacing salt-damaged bridges. The first step in the process was to remove the old surface with a jackhammer and chipping hammer.

In the process of hard-surfacing the state's road system, Iowa State Highway Commission engineers had established benchmarks in highway engineering. Anson Marston's desire for a state-managed highway system had come to fruition, although not always easily, and commission engineers and engineers at state universities had continued to

refine construction and maintenance methods and materials. From education to experimentation, Iowa's highway engineers continue to serve the public diligently under the Iowa Department of Transportation.

Bridge deck overlay, 1974
After the bridge deck was cleaned, workers poured a new concrete surface. The bridge could be reopened for traffic as soon as the concrete had cured.

Epilogue

On July 1, 1974, the Iowa State Highway Commission became part of the newly created Iowa Department of Transportation. The commission's 70-year tradition of excellence has continued under the new department as Iowa's highway engineers and researchers demonstrate ingenuity and dedication to their profession in the areas of safety, maintenance, and pavements. This quality public service is based on experimentation, sound judgment, and competition. In the coming century the demands of highway engineering will offer new challenges to Iowa's road builders. If the following examples from the past 20 years are any indication, Iowa's highway engineers will continue to perform in the best interests of the public.

The Iowa Highway Research Board continues to recommend funding for highway safety investigation. The Iowa State Engineering Research Institute, the contemporary Iowa State University engineering research agency, conducts many of these projects, as did its predecessor the Engineering Experiment Station. In 1981 professor R. L. Carstens reported for the institute on the effectiveness of rumble strips at intersections and grade crossings on primary and secondary roads.

Rumble strips, a series of grooves in the pavement, are employed on many Iowa roads to warn motorists of an impending stop sign at an intersection, and occasionally at railroad crossings. Carstens's

research was partially supported by the Iowa Highway Research Board. Some citizens were arguing for rumble strips on more Iowa roads to further reduce the number of accidents. Some county engineers, on the other hand, contended that rumble strips increase the number of accidents when bicyclists or moped riders swerve into oncoming traffic to avoid the strips.

Carstens systematically studied the effects of rumble strips and the frequency of accidents and identified situations where rumble strips should be used to improve highway safety. In this research, Carstens and R. Y. Woo, a graduate assistant, examined the accident experience at primary road intersections before and after the use of rumble strips and compared 88 secondary intersections with rumble strips to 88 similar intersections without rumble strips.

Rumble strips, 1968
Rumble strips warn motorists of approaching intersections. In 1982 the Iowa Highway Research Board and the Iowa State University Engineering Research Institute sponsored research to determine the best use of this warning device.

Crane, 1959 (facing page)
An early interstate project was I-80 west of Iowa City, Iowa.

Based on the results of the study rather than on anecdotal evidence, Carstens recommended guidelines for using rumble strips. Carstens concluded that in some instances rumble strips had little effect on the frequency of accidents. The study of primary roads indicated that locations with rates higher than two accidents per million entering vehicles benefited from rumble strips. If the accident rate at an intersection was higher than 2.5 accidents per million entering vehicles, the rate was always reduced by the use of rumble strips. The study found no indication of an increase in accidents because of the strips.[145]

Iowa engineers continued to pioneer concrete paving and pavement repair methods. Beginning in 1986 the process of "fast-track" concrete paving developed out of the Iowa method of bridge deck repair. Fast-track paving is a portland cement concrete paving technique that contains high levels of cement in the concrete mixture. Fast-track paving allows maintenance crews to overlay a new concrete surface on old pavement and open the road to traffic in less than 12 hours. Previously, resurfacing projects required a five- to ten-day curing period. Fast-track paving increases safety and convenience by reducing time that traffic lanes are closed or restricted.[146]

The fast-track process involves four steps: surface cleaning, joint identification, surface grouting, and placing and finishing the pavement. To clean the pavement workers use a process similar to sandblasting, but metal shot is used instead of sand. The blasting removes a thin layer of the old concrete, and the metal shot is reusable. The clean, dry surface allows a bonding grout and the wet overlay to penetrate the old pavement, creating a secure bond between the layers. After cleaning, joints in the old pavement are established and saw cuts are made to allow for expansion and contraction.

Tape is placed over existing cracks to prevent bonding and to reduce reflective cracking in the overlay at those points. The bonding grout, a thin mortar, is sprayed on the old surface, and the overlay is placed while the grout is wet. Local traffic can use the adjacent lane, and through traffic may pass in a few hours when the concrete has cured. The Iowa Department of Transportation developed this paving technique for road maintenance that offers a safer, more economical, and more convenient method of road repair.[147]

To maintain moderate paving and maintenance costs, the Iowa Department of Transportation encourages the use of asphalt and portland cement concrete paving for road work. Vernon Marks, contemporary research engineer at the department, says that "the Iowa DOT is able to build quality pavements with either asphalt concrete or portland cement concrete." The department and asphalt contractors have created asphalt concrete pavement mixes that do not rut and will bear the load of modern traffic. Portland cement concrete remains an important road building surface.

Because of these two strong industries, Iowa's interstate and highway system has performed well over the last 40 years. As former Iowa State Highway Commission engineer Robert Given says, "I think that it was a job well done with a lot of pressure. If you look at how [the highway system has] performed over the past twenty years, I think it's performed very well. The state roads have performed admirably, and some pavements have lasted more than fifty years."[148]

One area where problems continue to arise, however, is with pavement performance, as indicated in a study of Highway 20 near Fort Dodge in 1990. This investigation was related to cracking of portland cement concrete pavement only three years old. The limestone in the pavement had an excellent history of performance, and the cause of the degradation was unknown. When researchers from the Iowa Department of Transportation could not initially determine the cause of the pavement's failure, the agency hired a consultant to investigate the cracking.

The consultant reported that the cracking was a result of alkali-silica reactivity in the limestone aggregate. Alkali-silica reactivity results from a combination of sodium or potassium and silicon, which produces silica gel in the air, voids in concrete. To further investigate the degradation, the department and Iowa State University jointly purchased a scanning electron microscope and analyzed core samples removed from the pavement. The examinations revealed no silica gel concentrations but found "substantial amounts of sulfur, potassium, and aluminum, assumed to be ettringite." Ettringite will expand and then dissolve when subjected to road salt brine. This expansion is severe enough to create cracking in portland cement pavement. Currently the Iowa Department of Transportation and Iowa State University are investigating the deterioration further and encouraging other states to join in the research.

Iowa engineers continue to create and implement new procedures for highway engineering. Donna Buchwald, a transportation engineer for the Iowa Department of Transportation, recently helped develop an electronic record keeping system called FieldBook for field personnel to record progress on projects and monitor material stockpiles. The program works handily in the field on laptop computers. Buchwald remains involved in the American Association of State Highway and Transportation Officials (AASHTO) and its Construction Management System. The Iowa Department of Transportation is one of only three sites selected for the trial of the Construction Management System program. Buchwald's participation in an AASHTO advisory group contributed to Iowa's selection as a Construction Management System trial site.[149] Iowa continues to be a leader in research projects of national significance.

Many previous innovations developed by Iowa engineers also remain useful. The foamed asphalt technique developed by Ladis Csanyi in 1957, although used internationally, never entered common practice after 1970. However, the process has been refined by Soter International of Quebec. Soter International perfected the process and created equipment to pave roads in Canada, Mexico, and South Africa. The Soter company adapted an earlier improvement that eliminated the boiler in the process and produced a foamed asphalt that could be applied in all types of weather. The process is accepted by the Quebec Ministry of Transport for stabilization and paving.[150]

Iowa's highway engineers remain concerned with highway safety and economy. The lower speed limits established in the 1970s for four-lane divided highways with limited access were replaced in 1996 with limits similar to those of interstate highways, and pressure has been exerted to raise speed limits on other highways as well, a challenge to engineers to design safer roads or offer cautionary advice.

The rebuilding of and potential for additions to the federal-aid interstate system demand that the Iowa Department of Transportation investigate the most affordable highway designs to safely carry traffic at 65 or perhaps 75 miles an hour. Highway engineers reflect on the human cost of increased speed limits, and many believe the returns gained from increased legal speed limits are marginal compared to the human cost. As part of the re-evaluation of speed zones, the Iowa

In 1995 the national Transportation Research Board celebrated its 75th anniversary, and in 1996 the Federal-Aid Highway Act celebrated its 40th. Since the beginning of the century, construction of Iowa's road system has developed out of an interest in commerce, safety, and convenience and was made possible by significant research, machinery and procedures development, and educational efforts conducted by Iowa engineers. As the road system ages and increases in mileage, additional funds are needed for road work. In 1996 the state legislature appropriated $364 million, a record amount for highway work.

Flowable mortar
Iowa Department of Transportation engineers continue to develop new procedures for highway maintenance. Flowable mortar is a concrete slurry used to fill voids around restricted bridges that can be opened for public use.

Department of Transportation delivered a report to the Iowa legislature in 1996 linking traffic fatalities with faster speed limits. The perceived convenience from raised speed zones comes at an expense, according to the report, and is not necessarily the best public policy.[151]

Funding can be cyclical, but the service of Iowa's highway engineers is constantly in the public interest. Iowa's highway researchers and engineers remain dedicated to improving the Iowa road system and consequently the quality of life in the state. The citizens of Iowa can be confident that highway building in their state is based on a history of sound research, and that systematic, sensible research proceeds today under the direction of talented and farsighted men and women.

Mud jack, 1985
Innovations and practices developed by early Iowa State Highway Commission and Iowa State College engineers remain in use. Today Iowa's engineers continue to improve the road system for the state and the nation.

Notes

[1] Anson Marston, Charles Curtiss, and Thomas MacDonald, "The Good Roads Problem in Iowa," *Engineering Experiment Station Bulletin* 2 (June 1905): 4.

[2] *Annual Report of the Iowa State Highway Commission,* (Des Moines: The State of Iowa, 1919), 107. From this point forward citations for annual reports will list title, year, page.

[3] Raymond Kassell, interview with author, 19 January 1996; Iowa State Highway Commission Papers, Department of Civil Engineering Collection, University Archives, Parks Library, Iowa State University, Ames, IA.

[4] Iowa State Highway Commission, *Manual for Highway Officers, 1906 Revision* (Cedar Rapids, IA: The Republican Printing Company, 1906), 12.

[5] Marston, et al. "Good Roads Problem in Iowa," 7–10; Thomas R. Agg, *Construction of Roads and Pavements* (New York: McGraw-Hill Book Company, Inc., 1924), 103–104, 147.

[6] George Calvert, interview with author, 11 December 1995.

[7] Correspondence with F. E. Turneaure between 1 March 1904 and 30 April 1904. Marston elucidates more on these events in a letter to Elmina Wilson. Anson Marston to Elmina Wilson, 21 May 1904, Anson Marston Papers, University Archives, Parks Library, Iowa State University, Ames, IA.

[8] Anson Marston, "The State's Responsibility in Road Improvement," *Iowa Engineer* 7 (November 1907): 208–215.

[9] "Bridge Building Under The Old and New Systems," *Iowa State Highway Commission Service Bulletin* 2 (October 1914): 12; and a transcript of a speech by Marston to the delegates at the annual meeting of the Iowa League of Municipalities, "The Iowa State Highway Commission and Its Work," can be found in Herbert J. Gilkey, *Anson Marston: Iowa State University's First Dean of Engineering* (Ames, IA: Iowa State University Press, 1969), 146.

[10] *Civil Engineers Songbook*, no date, Department of Civil Engineering Papers, University Archives, Parks Library, Iowa State University, Ames, IA.

[11] For personnel see, *Annual Report of the Iowa State Highway Commission, 1913/1914*, 9. Verification of attendance at Iowa State College comes from University Archives, Parks Library, Iowa State University, Ames, IA.

[12] Anson Marston Papers, University Archives, Parks Library, Iowa State University, Ames, IA.

[13] A. B. Chattin and Ray McClure, "Good Roads Investigations," (Senior Thesis, Iowa State College, 1903), 1–15.

[14] L. T. Gaylord and T. H. MacDonald, "Iowa Good Roads Investigations" (Senior Thesis, Iowa State College, 1904), 86–90.

[15] Marston, et al., "The Good Roads Problem in Iowa," 4; For biographical information see Anson Marston Papers, University Archives, Parks Library, Iowa State University, Ames, IA.

[16] Anson Marston, et al., "Report of the Investigations on Drain Tile of Committee C-6," *Engineering Experiment Station Bulletin* 12 (April 1914); For examples of tests on materials by the Engineering Experiment Station see Thomas R. Agg, "Investigations of Gravel for Road Surfacing," *Engineering Experiment Station Bulletin* 15 (December 1916); Gilkey, *Anson Marston: Iowa State University's First Dean of Engineering*, 25–27; *Annual Report of the Iowa State Highway Commission, 1907/1908*, 15.

[17] Marston, et al., "The Good Roads Problem in Iowa," 4.

[18] Roy W. Crum, "Grading and Testing of Concrete Aggregate," *Iowa State Highway Commission Service Bulletin* 3 (February–March 1915): 10–11.

[19] *Annual Report of the Iowa State Highway Commission, 1906*, 12–20.

[20] For a discussion of systems in early twentieth century America see Alan Marcus and Howard Segal, *Technology in America: A Brief History* (San Diego: Harcourt, Brace, Jovanovich, 1989), 137–149.

[21] Shelby County, Iowa Engineer F. W. Sarvis stated that the road drag had "just about seen its day." F. W. Sarvis, "A General Description of the Work Done By Me As County Engineer of Shelby County in the Past Four Years" (Senior Thesis, Iowa State College, 1919), 7–8.

[22]*Rolfe Reveille,* 20 April 1905. The Iowa State Highway Commission published a *Manual for Iowa Road Officers* in 1905 which promoted the road drag. An analysis of the earthen road situation in Iowa can be found in the essay by George May, "The King Road Drag in Iowa," *Iowa Journal of History and Politics* 53 (July 1955): 247–272; *The Osceola Democrat,* 26 October 1905; *The Red Oak Express,* 27 October 1905.

[23]*Annual Report of the Iowa State Highway Commission, 1907/1908,* 6.

[24]The programs of the road schools are listed in Annual Reports of the Iowa State Highway Commission. Dates for the various schools were 29 December 1913 to 7 January 1914; 27 February to 1 March 1917; *Annual Report of the Iowa State Highway Commission, 1907/1908,* 14–17.

[25]*Annual Report of the Iowa State Highway Commission, 1913/1914,* 133–137.

[26]Iowa State Highway Commission, *Manual for Highway Officers, 1906 Revision.*

[27]"June Floods Worst in Nearly Forty Years, Wrought Havoc on Central Iowa Roads and Bridges," *Iowa State Highway Commission Service Bulletin* 6 (July 1918): 1–8. An article in the same issue relates the decision by a federal court of appeals that some of the Luten patents were invalid, see "Luten Patents Invalid, Says United States Court of Appeals," *Iowa State Highway Commission Service Bulletin* 6 (July 1918): 7; *Annual Report of the Iowa State Highway Commission, 1918,* 38–40.

[28]*Minutes of Iowa State Highway Commission,* 13 June 1913; Letter from Claude Coykendall to Fred White, 19 June 1913 and Report from Coykendall to White, addenda to *Minutes of Iowa State Highway Commission,* 21 June 1913.

[29]*Annual Report of the Iowa State Highway Commission, 1917,* 14; "Mahaska County Ice Gorge Shocks and Thrills Iowans with Series of Tragic Events," *Iowa State Highway Commission Service Bulletin* 4 (April 1916): 3–6, 15.

[30]"Iowa Leads Nation in Character and Quality of Her Bridges, says Lincoln Highway Official," *Iowa State Highway Commission Service Bulletin* 5 (November–December 1917): 12.

[31]*Annual Report of the Iowa State Highway Commission, 1915,* 12–13.

[32]Information on the Federal-Aid Road Act of 1916 can be found in Bruce Seeley, *Building the American Highway System,* (Philadelphia: Temple University Press, 1987), 44–51; for Iowa information see William Thompson, *Transportation in Iowa: A Historical Summary,* (Ames, IA: Iowa Department of Transportation, 1989), 73, 98–100; details of the use of federal funds prior to 1917 can be found in the *Annual Report of the Iowa State Highway Commission, 1916,* 35; "Winter Puts Stop to Army Cantonment and Federal Aid Brick and Concrete Road Building," *Iowa State Highway Commission Service Bulletin* 5 (November–December 1917), 7–8.; *Annual Report of the Iowa State Highway Commission, 1917,* 12; and *Annual Report of the Iowa State Highway Commission, 1918,* 4.

[33]Memorandum by T. H. MacDonald, April 7, 1919, Fred White Papers, University Archives, Parks Library, Iowa State University, Ames, IA.

[34]The Bergman Secondary Road Act of 1929 revised the secondary road laws for the state of Iowa. A principal change required the county board of supervisors to submit plans for secondary roads to the highway commission for approval.

[35]T. H. MacDonald, "Four Years of Road Building Under the Federal-Aid Act," *Public Roads* 3 (June 1920): 3–14.

[36]Fred White, "Iowa's Future Highway Policies," Fred White Papers, University Archives, Parks Library, Iowa State University, Ames, IA.

[37]"Six Year Improvement for Iowa Highways Proposed by Chief Engineer White," *Iowa State Highway Commission Service Bulletin,* (July–August 1923); "Three Year Road Building Program Outlined for Iowa Primary System," *Iowa State Highway Commission Service Bulletin* 16, (July, August, September 1925): 1–4.

[38]"Roads! Roads! Every Day for Twenty-Three Years! Enough For Any Man! Thinks Marston, Retiring," *Iowa State Highway Commission Service Bulletin* 15, (April, May, June 1927): 1–4; *Proceedings of the Highway Research Board,* 1935; Letter from Anson Marston to Fred White, Scrapbook. Fred White Papers, University Archives, Parks Library, Iowa State University, Ames, IA.

[39]"Tests Show That It Pays to Surface Dirt Roads When Average Daily Traffic Reaches 320 Tons," *Iowa State Highway Commission Service Bulletin* 12 (April, May, June 1924): 8–10; White's statements are reprinted in "Increased Federal Appropriations Imperative If Work Is To Be Completed In Ten Years," *American Highways* 3 (April 1924): 1–11.

[40]A transcript of Marston's speech appears in Anson Marston, "A National Program for Highway Research," *Good Roads* 58 (4 February 1920): 50, 62.

[41]Anson Marston, "A National Program for Highway Research," 50, 62.

[42]National Research Council, *Ideas and Action* (Washington, D.C.: National Research Council): 16–17, 280–283.

[43]Anson Marston, "The Theory of External Loads on Closed Conduits in Light of the Latest Experiments," *Proceedings of the Ninth Annual Meeting of the Highway Research Board,* 1929, 138–170.

[44]Examples of MacDonald's efforts are evident in the numerous articles he contributed to *American Highways* and *Public Roads.* For example, "Highway Recovery Accomplishments and Future Policies," *American Highways* 14 (January 1935): 5–8; and "Looking Toward the Highway Future," *American Highways* 16 (January 1937): 4–7.

[45]For the first three meetings held in 1921, 1922, and 1923, the board was organized under the National Research Council as the Advisory Board on Highway Research. For committees of 1922, see *Proceedings of the Second Annual Meeting of the Highway Research Board,* 1922, 8–14. *Proceedings of the Third Annual Meeting of the Highway Research Board,* 1922, 67, 97–99.

[46]Thomas H. MacDonald, "A Proposed Program of Highway Research," *Proceedings of the Ninth Annual Meeting of the Highway Research Board,* 1929, 24–28.

[47]Ibid., 29–31.

[48]"Two-Man Crew Spends Three Months Painting Center Line Mark for Accident Prevention," *Iowa State Highway Commission Service Bulletin* 14 (July, August, September 1926): 3–5, 15.

[49]W. H. Root, " Some New Developments In Road Construction and Maintenance," *Good Roads* 2 (February 1931): 42–46; W. H. Root, "Repair of Pavement Settlements," *American Highways* 10 (April 1931): 16–18.

[50]W. H. Root, "Repair of Pavement Settlements," *Proceedings of the Ninth Annual Meeting of the Highway Research Board,* 1929, 30–36; "Device for Raising Sunken Pavement," *Roads and Streets* 71 (March 1931): 118; W. H. Root, "Repairing Pavement Settlements," *Roads and Streets* 71 (April 1931): 123–125; "New Mud-Jack," *Roads and Streets* 77 (January 1934): 44.

[51]Root, "Repair of Pavement Settlements," 18.

[52]Clippings from newspapers across the state are in White's papers at Iowa State. *Ames Tribune,* 30 March 1933; *Dyersville Commercial,* 20 April 1933; *Hartley Sentinel,* 30 March 1933.

[53]A transcript of Fred White's testimony before the committee is in his personal papers; see Fred White Papers, University Archives, Parks Library, Iowa State University, Ames, IA.

[54]*Annual Report of the Iowa State Highway Commission, 1929,* 3.

[55]"Experimental Township Earth Road Built with Tractor Blade Grader Outfit Cost $146.50 Per Mile," *Iowa State Highway Commission Service Bulletin* 11 (October–November 1923): 3–7.

[56]Fred White, "Iowa a Pioneer in Batching by Weight and Use of Oversanded Gravel," *Roads and Streets* 11 (April 1927): 88–90.

[57]Roy Crum, "Method of Proportioning Concrete Materials—Screened and Unscreened Gravels." *Bulletin of the Engineering Experiment Station,* No. 60, May 1921.

[58]*Annual Report of the Iowa State Highway Commission, 1919,* 102–03.

[59]Biographical information for Bert Myers may be found in the files of the Library of the Iowa Department of Transportation.

[60]Roy Crum and Mark Morris, "Progress Report on Culvert Investigation," *Proceedings of the Fifth Annual Meeting of the Highway Research Board,* 1925, 271–291. Crum apparently delivered the paper, but Morris is listed as co-author in the *Proceedings.*

[61]*Annual Report of the Iowa State Highway Commission, 1935*, 16–17.

[62]Ibid., 17; Mark Morris, "Master Traffic Count on U.S. Highway 65, Near Ames, Iowa," *Proceedings of the Fifteenth Annual Meeting of the Highway Research Board*, 1935, 300–323.

[63]J. H. Griffith, "An Investigation of the Protective Values of Structural Steel Paints," *Bulletin of the Engineering Experiment Station,* No. 54, November 1919; T. R. Agg, "Traffic on Iowa Highways," *Bulletin of the Engineering Experiment Station,* No. 56, January 1920, 9.

[64]Agg's dates as a member of the Iowa State faculty are gathered from the Iowa State University Faculty Card File, University Archives, Parks Library, Iowa State University, Ames, IA.

[65]Thomas R. Agg, *Construction of Roads and Pavements* (New York: McGraw-Hill Book Co., 1924), 192–196.

[66]The members of the Highway Research Board executive committee and committee assignments are regularly listed at the front of all *Annual Proceedings,* e.g. "Highway Research Board, Officers and Members of Executive Committee," *Proceedings of the Fifteenth Annual Meeting of the Highway Research Board Held in Washington, D.C.,* Roy Crum, ed. (Baltimore: Waverly Press, 1936), 11. Herewith notes referring to Highway Research Board Proceedings will list author, title, *Proceedings of* meeting number *Annual Meeting of the Highway Research Board,* year, page number/s.

[67]Robley Winfrey, "Preliminary Studies Of The Actual Service Lives Of Pavements," *Proceedings of the Fifteenth Annual Meeting of the Highway Research Board*, 1935, 47–60.

[68]Ibid.

[69]Ibid.

[70]Robley Winfrey and Fred Farrell, "Life Characteristics of Surfaces Constructed on Primary Rural Highways," *Proceedings of the Twentieth Annual Meeting of the Highway Research Board*, 1940, 165–199.

[71]Ralph A. Moyer, "Skidding Characteristics of Road Surfaces," *Proceedings of the Thirteenth Annual Meeting of the Highway Research Board,* 1933, 123–168; Moyer, "Further Skidding Tests with Particular Reference to Curves," *Proceedings of the Fourteenth Annual Meeting of the Highway Research Board,* 1934, 123–130; an extensive bibliography of Moyer's work can be found on his vita, Ralph Moyer file, Department of Civil Engineering Personnel Files, University Archives, Parks Library, Iowa State University, Ames, IA; a transcript of a press release announcing Moyer's departure to California is also in his personnel file.

[72]Merlin Spangler, "Stresses in Concrete Pavement Slabs," *Proceedings of the Fifteenth Annual Meeting of the Highway Research Board,* 1935, 122–146.

[73]*Annual Report of the Iowa State Highway Commission, 1939,* 11–13; *Annual Report of the Iowa State Highway Commission, 1940,* 12.

[74]Iowa State Highway Commission, *Iowa Hiway Hilites,* May 1963, 29–36.

[75]"Removal of Lip Curb from Concrete Pavement," Report of the Iowa State Highway Commission, July 1948, Fred White Papers, University Archives, Parks Library, Iowa State University, Ames, IA.

[76]Ibid.

[77]"County Engineer Explains Paving Project Here With New Machine," *O'Brien County Bell,* 5 October 1949.

[78]"Slip-form Paving in the United States," Technical Bulletin 263, American Road Builders Association, 1967, 4–19, 74–75.

[79]*O'Brien County Bell*, 5 October 1949.

[80]Ibid.

[81]"Slip-form Paving in the United States," 4–19.

[82]Seeley, *Building the American Highway System,* 187–189.

[83]Boyden Sparks, "Our Highways Are Antiques," *Saturday Evening Post,* 15 January 1944, 16–18; Seeley, *Building the American Highway System,* 180–182.

[84]Moyer received the Highway Research Board Award for Ralph Moyer; "Motor Vehicle Operating Costs, Road Roughness and Slipperiness of Various Bituminous and Portland Cement Concrete Surfaces," *Proceedings of the Twenty-Second Annual Meeting of the Highway Research Board,* 1942, 13–52. Record of this award is listed in *Proceedings of the Twenty-Third Annual Meeting of the Highway Research Board,* 1943, 606.

[85]Merlin Spangler and H. O. Ustrud, "Wheel Load Stress Distributions Through Flexible Pavement Types," *Proceedings of the Twentieth Annual Meeting of the Highway Research Board,* 1940, 235–257; Merlin Spangler, "Wheel Load Stress Determination Through Flexible Type Pavements," *Proceedings of the Twenty-First Annual Meeting of the Highway Research Board,* 1941, 111.

[86]Merlin Spangler, "Wheel Load Stress Determination Through Flexible Type Pavement Types," 110–117; M. B. Russell and Merlin Spangler, "The Energy Concept of Soil Moisture and Mechanics of Unsaturated Flow," *Proceedings of the Twenty-First Annual Meeting of the Highway Research Board,* 1941, 435–450.

[87]Merlin Spangler, "The Structural Design of Flexible Pavements," *Proceedings of the Twenty-Second Annual Meeting of the Highway Research Board,* 1942, 199–224.

[88]Merlin Spangler, "Stresses and Deflections in Flexible Pipe Culverts," *Proceedings of the Twenty-Eighth Annual Meeting of the Highway Research Board,* 1948, 249–259.

[89]Merlin Spangler and O. H. Patel, "Modification of Gumbotil Soil by Lime and Portland Cement Admixtures," *Proceedings of the Twenty-Ninth Annual Meeting of the Highway Research Board,* 1949, 561–566.

[90]Merlin Spangler and Harry L. King, "Electrical Hardening of Clays Adjacent to Aluminum Friction Piles," *Proceedings of the Twenty-Ninth Annual Meeting of the Highway Research Board,* 1949, 589–599.

[91]Richard K. Frevert and Don Kirkham, "A Field Method for Measuring the Permeability of Soil Below a Water Table," *Proceedings of the Twenty-Eighth Annual Meeting of the Highway Research Board,* 1948, 433–442.

[92]Donald T. Davidson, "Exploratory Evaluation of Some Organic Cations as Soil Stabilizing Agents," *Proceedings of the Twenty-Ninth Annual Meeting of the Highway Research Board,* 1949, 531–536; and Donald T. Davidson and John Glab, "An Organic Compound as a Stabilizing Agent for Two Soil-Aggregate Mixtures," *Proceedings of the Twenty-Ninth Annual Meeting of the Highway Research Board,* 1949, 537–543.

[93]Robley Winfrey, "Kansas Highway Property Accounting Procedures," *Proceedings of the Twenty-Second Annual Meeting of the Highway Research Board,* 1942, 23–42.

[94]Ralph Moyer, "Motor Vehicle Operating Costs, Road Roughness and Slipperiness of Various Bituminous and Portland Cement Concrete Surfaces," *Proceedings of the Twenty-Second Annual Meeting of the Highway Research Board,* 1942, 13–52.

[95]Ting Ye Chu and Merlin Spangler, "Plant Stability Test for Hot-Mix Asphaltic Concrete Mixtures," *Proceedings of the Twenty-Ninth Annual Meeting of the Highway Research Board,* 1949, 159–167.

[96]Ralph Moyer, "Eleven-Year Study of the Expansion and Contraction of a Section of Concrete Pavement," *Proceedings of the Twenty-Fifth Annual Meeting of the Highway Research Board,* 1945, 71–81.

[97]Ralph Moyer, "Braking and Traction Tests on Ice, Snow, and on Bare Pavements," *Proceedings of the Twenty-Seventh Annual Meeting of the Highway Research Board,* 1947, 341–360.

[98]Ralph Moyer and D. S. Berry, "Marking Highway Curves With Safe Speed Indications," *Proceedings of the Twentieth Annual Meeting of the Highway Research Board,* 1940, 399–428.

[99]Ibid., 403. At the time of the study, Ohio had no state authority to establish speed limits.

[100]Vernon Gould, "Salvaging Old Pavements by Resurfacing," *Proceedings of the Twenty-Ninth Annual Meeting of the Highway Research Board,* 1949, 288–292.

[101]L. M. Clauson, "Secondary Road Surfacing Problems in Iowa," *Proceedings of the Twenty-Ninth Annual Meeting of the Highway Research Board,* 1949, 293–300.

[102]"List of Members, Iowa Highway Research Board," Highway Research Board Material, Department of Materials and Tests, Iowa Department of Transportation.

[103]*Annual Report of the Iowa State Highway Commission, 1955,* 27–30. Information on Mark Morris may be found in his personnel file in the Iowa Department of Transportation Library; Mark Morris Papers, Iowa Department of Transportation Library.

[104]Don McLean, interview with author, 12 December 1995.

[105]William Behrens, "Secondary Road Administration in Linn County, Iowa," *Proceedings of the Thirtieth Annual Meeting of the Highway Research Board,* 1951, 274–282.

[106]Elmer Clayton, "Slipform Pavement Shows Excellent Performance on Iowa County Roads," *Better Roads* 39 (April 1969): 24–26.

[107]George Norris, "An Era is Over," *Inside Magazine of the Iowa Department of Transportation,* May 1987, 5.

[108]Don McLean, interview with author.

[109]Ibid.

[110]Gerhard (Gus) Anderson, interview with author, 10 January 1996.

[111]Don McLean, interview with author.

[112]Colin Reuter, "They Said It Couldn't Be Done," *The Iowa Engineer* 59 (January 1959): 20–22.

[113]Ibid.

[114]"Motorist Aid Program," *American Highways* 51 (July 1972): 33; "Status of Iowa's HELP Program," *American Highways* 52 (April 1973): 29.

[115]Emmet M. Laursen and Arthur Toch, "Model Studies of Scour Around Bridge Piers and Abutments—Second Progress Report," *Proceedings of the Thirty-First Annual Meeting of the Highway Research Board,* 1952, 82–87.

[116]Philip Hubbard, "Field Measurement of Bridge-Pier Scour," *Proceedings of the Thirty-Fourth Annual Meeting of the Highway Research Board,* 1955, 184–188.

[117]Emmet Laursen, "Model-Prototype Comparison of Bridge-Pier Scour," *Proceedings of the Thirty-Fourth Annual Meeting of the Highway Research Board,* 1955, 188–193; D. E. Schneible, "Some Field Examples of Scour at Bridge Piers and Abutments," *Better Roads* 24 (August 1954): 21–24, 50–52.

[118]Csanyi's travels are documented in his personnel file as are articles and clippings related to his work on foamed asphalt. Ladis Csanyi Papers, University Archives, Parks Library, Iowa State University, Ames, IA.

[119]Robert M. Nady and Ladis H. Csanyi, "Use of Foamed Asphalt in Soil Stabilization," *Proceedings of the Thirty-Seventh Annual Meeting of the Highway Research Board,* 1958, 452–467.

[120]Iowa Department of Transportation, "Highways and Your Land," Publication 168, Iowa Department of Transportation, 1991.

[121]*Annual Report of the Iowa State Highway Commission, 1960,* 11–47.

[122]Gus Anderson, interview with author.

[123]Stanley Ring, interview with author, 7 December 1995.

[124]Don Anderson, interview with author, 19 December 1995.

[125]Ibid.

[126]Ibid.

[127]Ibid.

[128]Robert Given, interview with author, 3 March 1996.

[129]"Iowa and Federal Truck Size and Weight Laws: Historical Perspective," Report by the Office of Economic Analysis, Iowa Department of Transportation, January 1995, 1–6.

[130]Donald T. Davidson and T. Y. Chu, "Dispersion of Loess for Mechanical Analysis," *Proceedings of the Thirty-First Annual Meeting of the Highway Research Board,* 1952, 500–510.

[131]Donald T. Davidson and John B. Sheeler, "Clay Fraction in Engineering Soils: Influence of Amount on Properties," *Proceedings of the Thirty-First Annual Meeting of the Highway Research Board,* 1952, 558–563.

[132]F. B. Kardoush, James Hoover, and D. T. Davidson, "Stabilization of Loess with a Promising Quaternary Ammonium Chloride," *Proceedings of the Thirty-Sixth Annual Meeting of the Highway Research Board,* 1957, 736–754.

[133]J. B. Sheeler, J. C. Ogilvie, and D. T. Davidson, "Stabilization of Loess with Aniline-Furfural," *Proceedings of the Thirty-Sixth Annual Meeting of the Highway Research Board,* 1957, 755–772.

[134]L. W. Lu, and D. T. Davidson, et al., "The Calcium-Magnesium Ratio in Soil-Lime Stabilization," *Proceedings of the Thirty-Sixth Annual Meeting of the Highway Research Board,* 1957, 794–805

[135]Merlin Spangler, *Soil Engineering,* (Scranton, PA: International Textbook Service, 1951), vii.

[136]Ibid., 196–200; Merlin Spangler, *Soil Engineering,* 4th ed. (New York: Harper and Row, 1982).

[137]Chalmer J. Roy and Leo A. Thomas, et al., "Geologic Factors Related to Quality of Limestone Aggregates," *Proceedings of the Thirty-Fourth Annual Meeting of the Highway Research Board,* 1955, 400–412.

[138]Ibid.

[139]Don Anderson, interview with author.

[140]Otmar Zack, "Chemical Treatment of Stabilized Mineral-Aggregate Roadway Surfaces," *Proceedings of the Thirty-Fourth Annual Meeting of the Highway Research Board,* 1955, 431–433.

[141]James Johnson and W. G. Bester, "Widening and Resurfacing Highways with Concrete," *Proceedings of the Thirty-Fourth Annual Meeting of the Highway Research Board,* 1955, 434–438.

[142]Ibid.

[143]Melvin B. Larsen, "Iowa County Highway Maintenance Practices," *Proceedings of the Fortieth Annual Meeting of the Highway Research Board,* 1961, 497–511.

[144]Jerry V. Bergen and Berard C. Brown, "An Evaluation of Concrete Bridge Deck Surfacing in Iowa," Special Report of the Iowa Department of Transportation, April 1975, 1–15.

[145]R. L. Carstens and R. Y. Woo, "Warrants for Rumble Strips on Rural Highways," *Engineering Research Institute Project 1524,* (Ames, IA: Iowa State University, 1982), 42–45.

[146]Isabel Hendrickson, "Iowa Fast Track Concrete Paving," *AASHTO Quarterly,* October 1986, 65, 10–12

[147]Ibid.

[148]Vernon Marks, interview with author, 12 March 1996; Robert Given, interview with author.

[149]"Donna Buchwald Named Employee of the Year," Iowa Department of Transportation *Inside Magazine,* June 1996, 4

[150]"Foamed Asphalt Makes a Comeback," *Asphalt Contractor Magazine,* June 1995.

[151]Dena Gray-Fisher, "Report Confirms Speed Kills," Iowa Department of Transportation *Inside Magazine,* May 1996, 4.

Appendix

Time line of Iowa State Highway Commission and Iowa Department of Transportation Contributions to Highway Engineering

1905 Pipe Testing Program/ Concrete Culverts

Iowa may have been the first state to recognize the importance of concrete culvert pipe strength. As early as 1905 Anson Marston, dean of engineering at Iowa State College, developed tests and first proposed the "Imperfect Trench Method" for load theory on pipes.

1919 Proportioning Materials by Weight

This practice began as early as 1919 when the Iowa State Highway Commission persuaded two paving contractors to measure aggregate by weight instead of by volume. Roy Crum, Iowa State Highway Commission engineer of materials, developed this process. By 1922 it had become a required specification, and the method is used almost universally.

1920 Use of Impervious Film Under Concrete

The Iowa State Highway Commission was the first to use an impervious material, tar paper, under concrete slabs to retain moisture during the curing process.

1925 Use of Paper for Curing

The widely accepted practice of using paper for curing concrete began in Iowa in the early 1920s and continued until plastics became available and were used for that purpose.

1930 Mud Jack or Mud Pump

John Poulter, an Iowa State Highway Commission maintenance employee in Mt. Pleasant, developed the mud pump. The Iowa State Highway Commission first used the mud pump to raise sunken pavement in the Burlington area.

1934 24-Hour Traffic Counts for Classifying Traffic

The use of 24-hour traffic counts for classifying traffic was first practiced in Iowa, and it is now in use by many other highway engineers.

1935 Two-Point Loading of Test Beams

Iowa was first to use two-point loading of test beams for testing tensile strength of portland cement concrete beams.

1937 The Water/Alcohol Test

The water/alcohol test was developed in Iowa as a severe freezing-and-thawing test to determine the quality of aggregates.

1949 Slip-Form Paver

Perhaps the single most important highway-related invention to come out of Iowa, the slip-form paver was developed by employees of the Iowa State Highway Commission. Laboratory Chief James Johnson directed the development of the slip-form method of placing portland cement concrete pavement. This procedure laid highway pavements without the need for forms to support the vertical sides of the concrete. Slip-form paving is used nationally and in many foreign countries. Current slip-form pavers can place and finish over one mile of concrete per day.

1952 Tapered Inlet Culverts

The use of tapered or flared culvert inlets to increase hydraulic capacity was the result of a study supported by the Highway Research Board. Flared ends are used where feasible in Iowa. Other agencies and consultants have begun to implement these culverts.

1953 Cold Feed Calibrations

Iowa required that calibrations be made on cold feeds on hot mix plants. This led to elimination of the gradation unit, used to separate dried aggregate by size for separate proportioning into the mixer.

Highway surveyors, c. 1925 (facing page)
Surveyors, like these in northeast Iowa, were a common sight in the 1920s. Following the Primary Road Act of 1919, the state constructed primary roads connecting county seats and market towns.

1957 **Horizontal Cylinder Molds**
The use of horizontal cylinder molds for concrete compression tests originated in Iowa. This method was used to test prestressed concrete without the time lag necessary for capping. The cylinders went from the curing to the testing machine immediately. The horizontal mold controlled the surfaces so no capping was necessary. James Johnson, Iowa State Highway Commission laboratory chief, designed this method.

1958 **Aluminum I-Beam Bridge**
The world's first welded aluminum girder-type highway bridge was built over Interstate 35/80, northwest of the city of Des Moines. This aluminum bridge was built due to a delay in receiving steel but has since been replaced.

1959 **No Passing Zone Sign**
Iowa introduced the "no passing zone" pennant on US Highway 30. It has subsequently been adopted and included in the national *Manual on Uniform Traffic Control Devices for Streets and Highways*.

1961 **Machine Finishing Bridge Decks**
Iowa was certainly among the earliest, if not the first, to require machine finishing of bridge decks. W. W. Wickham presented a paper at an American Association of State Highway Officials meeting in Detroit on machine finishing. The process is now standard practice.

Prestressed Steel I-Beam Bridge
The first bridge with pre-tensioned steel beams was placed on US Highway 6 in Pottawattamie County. This bridge was built by placing camber in the beam when the cover plates were welded to the beam.

1962 **Photo/File Technique**
The Iowa State Highway Commission used Photo/File Technique on its 10,000-mile primary road system. This is a slow-moving film used to photograph the roadway. The prints are filed in the DOT library for future reference.

1963 **Iowa Method of Bridge Deck Repair and Resurfacing**
Iowa began experimental work now referred to as the "Iowa method" of bridge deck repair in 1963. A dense portland cement concrete with a low water-cement ratio was developed as a patching material. This mix is used for repairing small areas, or as a complete deck overlay.

1964 **Polyurethane Pavement Joints**
The use of polyurethane joints for pavement was developed in Iowa. The original use was for bridge approaches. The Iowa State Highway Commission worked jointly with Phelan (Midwest Manufacturing Company, Burlington, Iowa) in development of this method.

Full Depth, No Subbase Hot Mixed Asphaltic Concrete for Interstate
A section of Interstate 80 east of Iowa City built in 1964 is thought to be the first section of interstate highway with hot mixed asphaltic concrete as the full depth of the pavement structure, resting on earth with no lower base or subbase.

1966 **Continuous Reinforced Paving/No Transverse Bars**
This method pioneered the development of machinery to place reinforcing steel without chairs (cross frames) to support it. The method eliminated a great amount of hand labor and saved time in paving interstate highways.

1968 **Optimum Enforcement Level**
Iowa is the only state that has made a study by research consultants to determine the "Optimum Enforcement Level" for Traffic Weight Operations.

1971 **First in Miles of Secondary Road Portland Cement Concrete**
As of June 1, 1971, Iowa led the nation in the number of miles of portland cement concrete paving in its secondary system. About 2,750 miles of portland cement concrete highways have been constructed or are under construction in Iowa.

1976 Recycling of Portland Cement Concrete

In 1976 Iowa recycled its first portland cement concrete roadway. Good aggregates are valued and in increasingly short supply, and transportation costs are prohibitive. Recycling permits the reuse of aggregate at a reduced cost. Portland cement concrete recycling is now another design alternative when reconstruction is considered.

1985 Flowable Mortar

In 1985 Iowa constructed its first project using flowable mortar as fill material. The nation's road system contains thousands of bridges that are decrepit and restricted to low weight vehicles. This innovative method of supporting older bridges with a minimum inconvenience to the traveling public is an excellent alternative to many bridge replacement projects.

Thin Bonded Overlays on Grade

By using the same general method and principle used in the Iowa method of bridge deck repair, Iowa developed an overlay method using low slump concrete and a cement grout bonding agent. The overlay on lower traffic roads (US 20), on secondary roads, and on Interstate 80 in Pottawattamie County were highly successful. Other states, counties, and cities are overlaying hundreds of miles of pavement annually.

Drum Mixer Asphaltic Concrete Plant

By mixing the asphaltic cement concrete in low profile drum-dryers and elevating the mix into overhead storage bins, these plants are more energy efficient, require less setup and takedown time, and are easily transported. This continuous drying-mixing process is more efficient than the older conventional batching plant setup.

Bell Jointed Culverts

The use of bell jointed culverts in Iowa is a possible first.

Selected Bibliography

Agg, Thomas R. *Construction of Roads and Pavements.* New York: McGraw-Hill Book Company, Inc., 1924.

Annual Report of the Iowa State Highway Commission, 1904–1974.

Chattin, A. B. and Ray McClure. "Good Roads Investigations." Senior Thesis, Iowa State College, Ames, IA, 1903.

Crum, Roy W. "Grading and Testing of Concrete Aggregate." *Iowa State Highway Commission Service Bulletin* 3 (February–March 1915): 10–11.

Csanyi, Ladis. Papers. University Archives, Parks Library, Iowa State University, Ames, IA.

Davis, Rodney O. "Iowa Farm Opinion and the Good Road Movement, 1903–1904." *Annals of Iowa* 37 (Summer 1964): 321–338.

Gaylord, L. T. and T. H. MacDonald. "Iowa Good Roads Investigations." Senior Thesis, Iowa State College, Ames, IA, 1904.

Gilkey, Herbert J. *Anson Marston: Iowa State University's First Dean of Engineering.* Ames, IA: Iowa State University Press, 1969.

Iowa State Highway Commission. *Manual for Highway Officers, 1906 Revision.* Cedar Rapids, IA: The Republican Printing Company, 1906.

_____. "Fifty Years of Highway Progress." *Iowa Hiway Hilites,* May 1963.

"June Floods Worst in Nearly Forty Years, Wrought Havoc on Central Iowa Roads and Bridges." *Iowa State Highway Commission Service Bulletin* 6 (July 1918): 1–8.

MacDonald, T. H. "Four Years of Road Building Under the Federal-Aid Act." *Public Roads* 3 (June 1920): 3–14.

_____. "A Proposed Program of Highway Research." In *The Proceedings of the Ninth Annual Meeting of the Highway Research Board,* edited by Roy Crum. Baltimore: Lord Baltimore Press, 1930, 24–28.

"Mahaska County Ice Gorge Shocks and Thrills Iowans with Series of Tragic Events." *Iowa State Highway Commission Service Bulletin* 4 (April 1916): 3–6, 15.

Marston, Anson, Charles Curtiss, and Thomas H. MacDonald. "The Good Roads Problem in Iowa." *Engineering Experiment Station Bulletin* 2 (June 1905): 1–25.

Marston, Anson. "The State's Responsibility in Road Improvement." *Iowa Engineer* 7 (November 1907): 208–215.

_____. "A National Program for Highway Research." *Good Roads* 58 (4 February 1920): 50–62.

_____. Papers. University Archives, Parks Library, Iowa State University, Ames, IA.

May, George. "The King Road Drag in Iowa." *Iowa Journal of History and Politics* 53 (July 1955): 247–272.

_____. "The Good Roads Movement in Iowa." *Palimpsest* 36 (January 1955): 1–64. Reprinted in *Palimpsest* 46 (February 1965): 65–122.

_____. "Post War Road Problems." *Palimpsest* 46 (February 1965): 116–128.

National Research Council. *Ideas and Action: A History of the Highway Research Board.* Washington, D. C.: National Research Council, 1971.

Norris, George. "An Era is Over." Iowa Department of Transportation *Inside Magazine,* May 1987, 5.

Proceedings of the Highway Research Board, 1921–1974.

Root, W. H. "Repairing Pavement Settlements." *Roads and Streets* 71 (April 1931): 123–125.

Sarvis, F. W. "A General Description of the Work Done By Me As County Engineer of Shelby County in the Past Four Years." Senior Thesis, Iowa State College, Ames, IA, 1919.

Seeley, Bruce. *Building the American Highway System.* Philadelphia: Temple University Press, 1987.

Slip-form paver, c. 1950 (facing page)
Iowa's invention of the slip-form paver was a pivotal contribution to highway construction.

"Six Year Improvement for Iowa Highways Proposed by Chief Engineer White." *Iowa State Highway Commission Service Bulletin* 11 (July–August 1923): 1.

"Slip-form Paving in the United States." Technical Bulletin 263, American Road Builders Association, 1967.

Spangler, Merlin. *Soil Engineering*. Scranton, PA: International Textbook Service, 1951.

Thompson, William. *Transportation in Iowa: A Historical Summary*. Ames, IA: Iowa Department of Transportation, 1989.

"Three Year Road Building Program Outlined for Iowa Primary System." *Iowa State Highway Commission Service Bulletin* 13 (July, August, September 1925): 1–4.

White, Fred. Papers. University Archives, Parks Library, Iowa State University, Ames, IA.

"Winter Puts Stop to Army Cantonment and Federal Aid Brick and Concrete Road Building." *Iowa State Highway Commission Service Bulletin* 5 (November–December 1917), 7–8.

Index